ROUTLEDGE LIBRARY EDITIONS: NUCLEAR SECURITY

Volume 23

AUSTRALIA AND NUCLEAR WAR

AUSTRALIA AND NUCLEAR WAR

Edited by
MICHAEL DENBOROUGH

Routledge
Taylor & Francis Group
LONDON AND NEW YORK

First published in 1983 by Croom Helm Australia

This edition first published in 2021
by Routledge
2 Park Square, Milton Park, Abingdon, Oxon OX14 4RN

and by Routledge
52 Vanderbilt Avenue, New York, NY 10017

Routledge is an imprint of the Taylor & Francis Group, an informa business

British Library Cataloguing in Publication Data
A catalogue record for this book is available from the British Library

ISBN: 978-0-367-50682-7 (Set)
ISBN: 978-1-00-309763-1 (Set) (ebk)
ISBN: 978-0-367-52938-3 (Volume 23) (hbk)
ISBN: 978-1-00-308013-8 (Volume 23) (ebk)

Publisher's Note
The publisher has gone to great lengths to ensure the quality of this reprint but points out that some imperfections in the original copies may be apparent.

Disclaimer
The publisher has made every effort to trace copyright holders and would welcome correspondence from those they have been unable to trace.

AUSTRALIA AND NUCLEAR WAR

Edited by Michael Denborough

CROOM HELM AUSTRALIA

Croom Helm Australia Pty Ltd, 28 Kembla Street, Fyshwick, ACT 2609
Croom Helm Ltd, Provident House, Burrell Row, Beckenham, Kent BR3 1AT

Australian Library Cataloguing-in-Publication Data

Australia and nuclear war

Includes bibliographical references and index
ISBN 0 949614 05 X
ISBN 0 949614 09 2 (pbk.)

1. Atomic warfare. 2. Australia — National security. I. Denborough, M. A. (Michael Antony), 1929-

355'.0217

Typeset and Printed by Southwood Press, Sydney

To the children of the world

All royalties are to be paid to the Australian National Universitv Fund for the Prevention of Nuclear War.

CONTENTS

ACKNOWLEDGEMENTS

Planning for the symposium which led to this book, began in April 1982. The Steering Committee members were: Professor Frank Barnaby, Mrs Wendy Barnaby, Mrs Maureen Barnett, Dr Stephen Boyden, Dr H. C. Coombs, Dr Michael Denborough (Chairman), Ms Jane Kriegel, Mr Hugh Littlewood, Mrs Suzanne Ridley, Dr Hugh Saddler, Professor Ralph Slatyer, Mr David Ingle Smith, Professor G. H. Taylor and Mr Robyn Williams.

From January 1983, the detailed organisation of the symposium, which was entitled 'The Consequences of Nuclear War for Australia and its Region' was carried out by the Centre for Continuing Education at the Australian National University. Those who were involved included Mrs Dorothy Davis, Dr Michael Denborough, Dr Chris Duke, Ms Jane Kriegel, Ms Ann Mackay, Mr Brendan O'Dwyer, Ms Judy Pearce, Mr Colin Plowman and Mrs Beth Stoodley. Mrs Maureen Barnett, Co-ordinator of University Information, Mr John Dash and Mr Sean Whelan publicised the symposium, which was held at the Australian National University on May 30 and 31, 1983.

I am indebted to Mrs Stella Cornelius for her advice in the organisation of the symposium. Mr Nicholas Williams, Mrs Suzanne Ridley, Mrs Erica Denborough and Mrs Leigh van Haalen were very helpful in preparing the book. Mr Stephen Cole carried out the artwork.

Michael Denborough

NOTES ON CONTRIBUTORS

Desmond Ball is Senior Fellow at the Strategic and Defence Studies Centre, Australian National University.

Frank Barnaby is Professor of Peace Studies at the Free University of Amsterdam.

H. C. Coombs is Visiting Fellow at the Centre for Resource and Environmental Studies, Australian National University.

P. J. Crutzen is Director of the Max Planck Institute for Chemistry, Mainz, West Germany.

Michael Denborough is Professorial Fellow at the Department of Medicine and Clinical Science, John Curtin School of Medical Research, Australian National University.

Bernard Feld is Professor of Physics at Massachussetts Institute of Technology.

I. E. Galbally is Principal Research Scientist at the CSIRO Division of Atmospheric Research.

Oleg Gavrilov is Chairman of the Medical Council, Ministry of Health, USSR.

David Hackworth is the most decorated soldier in US military history, but has now left the army and lives in Australia.

Judy Irving is an independent American film-maker.

John Langmore is Economic Adviser to the Treasurer of Australia.

R. H. Mathams is Former Director of Scientific and Technical Intelligence for the Joint Intelligence Organisation.

Heather Ogilvie is from the Sydney Filmmakers Co-Operative.

Barrie Pittock is Principal Research Scientist at the CSIRO Division of Atmospheric Research.

H. Rodhe is Professor of Chemical Meteorology at the University of Stockholm.

Susan Ryan is Minister for Education and Youth Affairs and Minister Assisting the Prime Minister on the Status of Women.

Nancy Shelley is a Canberra peace activist.

Roland Vogt is a Member of Parliament in the Federal Republic of Germany representing the Green Party.

J. A. Ward is Secretary of the Australian branch of the Medical Association for Prevention of War.

Patrick White is a Nobel Prize winning author.

PREFACE

This book examines the possibility of nuclear war, its implications in human terms and strategies for avoiding it. It reflects the desire of men and women to mobilise their intellectual efforts to come to grips with the massive problems involved. We are considering the very survival of the human race and the civilisation built up over several millennia. Nations and cultures have recovered from conventional war — even from what we used to call 'total war' — in a comparatively short time, but the consequences of nuclear war may well be irreversible.

This is an emotional issue. A university should not avoid issues simply because they arouse strong emotions. Nevertheless, the task of a university is not to dwell on the emotional content of an issue — we have no special expertise in that. What we can do is subject the issue itself and its diverse components to dispassionate and objectively critical analysis. By these means, we may learn to develop strategies for averting nuclear disaster. I hope that the Australian National University has made a significant contribution to this all important cause by arranging the symposium which has led to this book.

Emeritus Professor Peter Karmel
Vice-Chancellor, Australian National University

INTRODUCTION

Planet Earth is galloping towards destruction due to our irrationality. The threat comes from the spiralling nuclear arms race and the ever increasing danger of nuclear war, but, lemming-like, we are doing very little to correct this. In Australia, it is remarkable that there is very little discussion about what should be done about this issue, which was not even mentioned by either of the two major political parties in the Federal Election this year.

The purpose of this book is to examine nuclear war from an Australian point of view, with the aim of encouraging debate in our community about the most important problem that the world has ever had to face. This debate is essential if we are to devise individual, collective and national strategies to prevent nuclear catastrophe.

The book is arranged in three parts. The first considers the nuclear arms race, the second the consequences of nuclear war, and the third is concerned with the prevention of nuclear war.

The first chapter examines the likelihood of nuclear war, and it is appropriate that it should be written by Frank Barnaby, who, over the last twelve years, has done more than anyone to alert us to the dangers of the nuclear arms race. Barnaby helps us to understand the awesome nature of the superpower nuclear arms race and the causes of it. The statistics which he quotes are almost unbelievable, for instance the superpowers have nuclear warheads each of which contains as much explosive power as all the explosives used in war so far put together. The nuclear arsenals are now, and have for many years been, much larger than needed for any conceivable military, political or strategic reason. Whilst this situation is bad enough in itself, Barnaby feels that it is rapidly getting worse with the development and deployment of more accurate and reliable nuclear weapons. This is leading strategists in the superpowers to think in terms of fighting or winning a nuclear war, which considerably increases the risk of a nuclear world war breaking out. Vast bureaucracies have grown up in the superpowers to form giant academic-bureaucratic-military-industrial complexes which are out of political control.

The nuclear arms race was discussed at the symposium by Julie Dahlitz, who has had considerable experience working for nuclear disarmament in the United Nations. She also considers that in 1983 there is a world strategic crisis, which increases the risk of nuclear war. Dr Dahlitz describes three distinct facets to this crisis. These are a lack of strategic stability due to developments towards a disabling first strike capability, an abandonment of the conditions needed for crisis stability due to a lack of confidence in the ability to detect early warning of attack from the new weapons, and a lack of compromise facility because it is not feasible to verify adherence to the terms of nuclear weapons agreements. Particularly dangerous developments are proposals to deploy cruise missiles in Europe as these are difficult to detect, and to militarise outer space.

The first chapter is concerned with global strategies, but the second, written by two prominent Australian defence strategists, Desmond Ball and R. H. Mathams, considers the consequences of nuclear war for Australia. These authors clearly make the important point that the United States bases in Australia are vital for United States nuclear strategies, and would almost certainly be attacked in the event of a nuclear conflict between the US and the USSR. Other targets in Australia would be the naval base at Cockburn Sound (WA) and the RAAF base at Darwin (NT), both of which are visited by US nuclear forces.

John Langmore points out that nuclear weapons are strategically, bureaucratically and politically integrated with conventional and other forces. He then goes on to consider the effects of military expenditure on the current international economic crisis. About half of world military expenditure is by the superpowers, and these together with West Germany, the United Kingdom, France and Saudi Arabia, account for three quarters of all military expenditure. Langmore makes the point that military expenditure is luxury consumption: every dollar spent on military activity is a dollar which cannot be spent elsewhere. As military goods and services have little economic usefulness, either for consumption or for further production, military expenditure is unproductive. It is not aimed at increasing productivity or growth and instead diverts financial and human resources from productive activity. Examples of this are President Reagan's cuts in education, health, income security and overseas aid programmes to make way for defence. This has had a strong impact in Australia, as high US interest rates have contributed to high Australian interest rates, which have been one of the causes for the contraction in Australian private investment, and for the fall in net business profits and housing construction. The arms race is exacerbating economic and social problems. It is reducing living standards, and increasing the rate of inflation and the number of people who are unemployed.

Whilst the problems caused by the nuclear arms race are serious enough, the outbreak of nuclear war is definitely contraindicated medically. The consequences would be so serious that the medical profession refers to it as 'The Last Epidemic'. If a nuclear war broke out between the US and the USSR, in which less than half of the total explosive power in the Soviet and the American arsenals was used, 750 million out of a total urban population of nearly 1.3 billion people in the Northern Hemisphere would be killed outright. Most of these would either be crushed or burnt to death. Some 340 million would be seriously injured. Of the remaining 200 million, many would perish from the latent effects of radiation and from epidemics such as cholera, typhoid, infectious hepatitis or dysentery. The living would almost certainly envy the dead. It is estimated that 200 nuclear weapons are targeted on the United Kingdom. A single one of these exploded over St Paul's Cathedral in London would totally disrupt the medical services of the United Kingdom.

John Ward, who is Secretary of the Australian Branch of the Medical Association for Prevention of War, feels that if a nuclear war occurred between the US and the USSR, not only would the three US bases, Cockburn Sound and Darwin be attacked, but so also would be the port areas of Sydney, Wollongong, Newcastle, Melbourne, Geelong, Adelaide and Fremantle. The deaths and injuries from such an attack would be an unimaginable catastrophe. If an attack is made on only the US bases and the Australian bases used by US nuclear forces, there would be almost

300 000 deaths, but if the port areas are attacked as well, the deaths would number about 4 000 000. There is obviously no possibility of a medical response to such an event.

Dark Circle is a remarkable film about the nuclear age, and it illustrates clearly that there are already very significant medical consequences from nuclear power plants and the making of nuclear weapons without even considering a nuclear war. One of the two filmmakers, Judy Irving, writes about the film with Heather Ogilvie. The film provides a very dramatic and moving picture of how the building, testing and selling of hydrogen bombs has led to the contamination of the environment with plutonium, and how this has resulted in cancer and death in unsuspecting American citizens.

Dr H. C. Coombs, the distinguished academic and public servant, points out that the Australian economy is closely integrated with the rest of the world, particularly with the US, Europe and Japan. The indirect effects alone of a nuclear conflict in the Northern Hemisphere would confront Australia with international and domestic economic and organisational problems, besides which our current difficulties would pale into insignificance. During a superpower conflict, Australia's importance to its allies would be marginal. If there were a direct nuclear attack on Australia, radioactive fallout would have deleterious effects on food and other agricultural products both for consumption and for export. Should large population centres be destroyed, the social and economic dimensions of the catastrophes are hard to conceive, let alone to define.

Barrie Pittock has carried out a detailed and careful scientific review of the atmospheric consequences of nuclear war. He cautions that definitive conclusions about the atmospheric effects of a major nuclear war cannot be made before the event, because a number of basic uncertainties exist. Many serious consequences are described. These include contamination of the upper atmosphere or stratosphere, whence the contaminants would be transported from the Northern to the Southern Hemisphere over one or two years. There are three possible sources of contamination of the stratosphere from nuclear war. One is from detonations either in or above the stratosphere, and a second is from powerful nuclear explosions in the lower atmosphere in which the fireballs penetrate the stratosphere due to the energy of the explosions themselves. The third source is from major urban fires initiated by nuclear explosions which by themselves would not be energetic enough to penetrate the stratosphere. Several atmospheric chemical processes could be seriously effected by nuclear war. One of these would be a depletion of the ozone layer of the stratosphere leading to an increase in biologically damaging ultraviolet radiation. This would lead to sunburn, skin cancers and damage to the cornea of the eye, leading to cataracts and blindness in all animal life. There would also be serious climatic effects.

Galbally, Crutzen and Rodhe have considered the changes in the atmosphere that may occur over Australia after a nuclear war in the Northern Hemisphere. Their greatest concern is the possibility of immediate particle production, which could be two orders of magnitude more than the natural loading of the atmosphere. Huge black clouds could be formed in the mid-latitudes of the Northern Hemisphere absorbing and reflecting all the incoming solar radiation. No firm prediction of the subsequent behaviour of these clouds can be made. It is possible that they could pass into the stratosphere and spread globally, and that they could cause substantial changes in the surface temperature of the earth. These clouds could persist for at least one or two weeks and, depending on changes in atmospheric behaviour, perhaps for much longer.

By the end of the second part of this book, it is clear that the mere possession of nuclear weapons is irrational, and that to contemplate the development of new ones is madness. And yet in 1983 President Reagan plans to spend $1.6 trillion ($1 600 000 000 000) on expanding the so-called defence programme, and this includes plans to put nuclear warfare into space. Anything more terrifying than this is difficult to imagine. The US Secretary for Defence, Mr Casper Weinberger, says that it is essential, after losing the Vietnam War, that the US should win the next war, whereas it should be obvious that any nuclear war will be unwinnable. The strategists in the Pentagon are making crazy plans to ensure that US missiles are not destroyed by a Soviet nuclear strike. These plans include a series of underground tunnels so that the US missiles can be continually on the move, and deployment of the missiles on barges on the nation's waterways, or in the air in giant balloons. Although the world hears little about the Soviet defence plans, the strategists in the USSR and the US seem to speak the same language, and to use similar excuses for the ever-expanding arms race. In this language, the safety of US and USSR nuclear missiles is far more important than the lives of US and

USSR citizens. These recent developments reinforce Frank Barnaby's view that the nuclear arms race is out of political control, and that the only hope for survival of the human race lies with the mobilisation of public opinion of the ordinary people of this world. Only in this way can sufficient pressure be exerted on political leaders to overcome the influence of the military-industrial-academic-bureaucratic complex. And the amount of time that is left is short.

The third part of the book is concerned with examples of how this public opinion can be mobilised, so that nuclear war can be prevented.

The first chapter in this part is written by Bernard Feld, who is very well qualified to write on the role of the scientist. As a young physicist, Feld was guilty of 'the original sin', as he describes it, being part of the team which made the first nuclear weapons. Unlike some of the others involved, however, Feld has been working against nuclear weapons since then, as an editor of *The Bulletin of Atomic Scientists,* in Pugwash, and in many other roles. Feld points out that it requires only a low level of technical competence to fashion a crude nuclear bomb of devastating power, and he is concerned that it will not be very long before terrorist groups have acquired this capability. He is concerned that the acquisition of high-power nuclear reactors by many countries will accelerate the spread of nuclear weapons. Feld explains that scientists and engineers have an important role to play in educating both governments and the public in their respective nations, on the vital importance of nuclear arms control, and on the relevance and value of particular actions under consideration.

Academician Professor Oleg Gavrilov describes the important role that the medical profession is playing in the prevention of nuclear war. He states that the possibility of nuclear war is the greatest existing threat to mental health. He explains the discrepancy between the very real threat of a nuclear catastrophe and the passive attitude to this massive problem by an 'exclusion' phenomenon. This refers to the human desire to exclude from consciousness any unpleasant thoughts about possible cancer, inevitable death and so on. Gavrilov explains that the more severe the threat to humans, the greater is the abnormal behavioural response. Albert Einstein said 'We shall require a substantially new manner of thinking if mankind is to survive.' Gavrilov says that this manner of thinking can only be based on truth, courage and honesty, and believes that it is these criteria that have brought tens of thousands of physicians together to prevent nuclear war. These physicians are dedicated to the cause of preserving life and health in various parts of the world regardless of nationality, and religious or political views. He refers to the Society of International Physicians for the Prevention of Nuclear War which was founded by a US and a Soviet physician in 1981 and has been growing extremely rapidly since then. Gavrilov believes that patients who trust doctors with their health and life, must be informed about the realities of the present nuclear danger. Soviet physicians have published a book, *The Danger of Nuclear War* and the views of the Soviet and US physicians are widely discussed on Soviet TV and radio, in the press and in special medical journals.

David Hackworth, the most decorated US soldier in history, illustrates the important role that the military can play. Disillusioned with US nuclear war policies, he migrated to Australia three years ago, and is now working for nuclear disarmament. Hackworth has spent most of his life dealing in military solutions, and knows first-hand the leaders in the US who are saying that nuclear war is winnable, survivable and manageable. He believes that the US has become a militarised nation, and says that many key positions, previously chaired by civilians who thought in terms other than military solutions, are now filled by generals. This is a matter for grave concern when we consider that since 1945, the use of nuclear weapons has been recommended on 20 occasions by US generals. Hackworth says that we must change our ingrained habits and assumptions as a matter of urgency, and convince our political leaders that the nuclear problem transcends all other issues. Because of our alliance with the US and the presence of US military bases in our country, Hackworth is convinced that Australia could lead the way to world nuclear disarmament, but stresses that we must stand on our own feet, and take a fresh look at the issue.

The Green Party in the Federal Republic of Germany is of considerable importance because it provides an example of how individuals who are concerned about the dangers of nuclear war, can translate their anxiety into political reality. One of the two founders of the Green Party was Roland Vogt, and he describes how it came about. Vogt first met Petra Kelly at an anti-nuclear power plant demonstration in Wyhl in 1975 and they became firm friends. It is heartening to learn that Australians played a part in the formation of the Green Party, following the visit of Petra Kelly to Australia in 1976. When she returned to Germany she was able to bring encouragement to the anti-nuclear movement in her own country from developments which had occurred in Australia. These included a Green Ban against the transportation of uranium, a movement to 'Keep Uranium in the Ground', and the development of co-operation between the Anti-Uranium Movement, some trade unions and Aboriginals. In 1976, Kelly and Vogt founded the Green Party which combined opposition to nuclear power plants and to nuclear weapons. They argued that it was illogical to fight only against a dozen nuclear power plants and to ignore 6000 atomic warheads on German soil. The ecological and peace movements in West Germany joined forces and in 1983, 27 members of the Green Party were elected to the Bundestag (the Parliament in the Federal Republic of Germany). Their main aim at the moment is to break the armament cycle by preventing the deployment of Pershing-II and cruise missiles in West Germany. They have demonstrated recently in East Berlin with the East German autonomous Peace Movement. The Green Party is very much in favour of Social Defence. Vogt strongly supports the idea of an International Summit for Survival co-chaired by the Australian government. He asks for Australian financial support for the European Peace Movement which urgently needs it for opposition to the installation of 112 cruise missiles in South Sicily, for the women of Greenham Common, and for other affinity groups which are organising the non-violent blockade of proposed sites for Pershing-II and cruise missiles in Europe. He also urges Australians not to allow their country to be used by a superpower as a military colony.

Senator Susan Ryan makes an important statement about the commitment of the Australian Government to nuclear disarmament. The Australian Government recognises the withering of détente between the two great powers and the growing tension between them, and the growing danger of a nuclear catastrophe. As host to important US nuclear weapons facilities and as an ally of the US, the Australian Government intends to express its views forcibly about nuclear disarmament, in international forums and in bilateral discussions. The Australian Government will urge the leadership of the two superpowers to resume control of the technology of nuclear war, to abandon the doctrine of limited nuclear war and of winning a nuclear contest, and to urge the development of a genuine dialogue with a readiness to find accommodation and to accept restraints. The Australian Government has appointed an Ambassador for Disarmament to represent Australia in all international forums on disarmament and arms control.

Nancy Shelley makes a compelling case to support the view that the nuclear threat which we are now facing is due to the narrow, unimaginative and arrogant thinking of male scientists and technocrats. Women throughout the world have good cause not to trust male judgement anymore. Women feel that men have missed their chance to bring peace. Women now wish to stress the real priorities which are those of life, not of fear, of co-operation and not competition, and of human needs rather than weapons construction. Nancy Shelley points out the close correlation between sexism and militarism, and gives many striking examples of the actions of women throughout the world to promote peace. This is always non-violent because of a commitment to bringing about a non-violent world. The Shibokusa women have long been resisting militarism at the foot of Mount Fuji in Japan. Scandinavian women have arranged marches from Copenhagen to Paris in 1981, and to Moscow and Minsk in 1982. Other examples are cited from America, Australia, the Argentine, Holland, Italy and, of course, the heroic stand at Greenham Common in England.

For the last chapter, Patrick White, the Nobel Prize-winning author, has written a literary masterpiece. Patrick White urges his fellow Australians to acquire identities of their own, and to practise non-attachment which entails the practice of all the virtues — most important charity, but also that of courage, and the cultivation of intelligence, generosity and disinterestedness. He asks each of us to search for the good faith in us which may help save the world, even if we risk turning ourselves into outsiders in this materialistic Australian society. Patrick White gives numerous examples of how non-violent resistance can achieve positive results, and inspires us to rouse ourselves to do all in our power to prevent the children we have created from suffering a fate similar to that thrust upon the children of Hiroshima and Nagasaki.

With the exception of *Dark Circle,* the chapters in the book are versions of papers given at a symposium on nuclear war held at the Australian National University. On the afternoon of the last day of the symposium, it was decided to send letters of protest to the leaders of the countries most involved in the nuclear arms race.

Letters were sent to President Reagan, Chairman Andropov, Prime Minister Thatcher, President Mitterrand, Prime Minister Hawke, Chairman Ye Jianying, Chancellor Kohl, Prime Minister Lubbers, President Pertini, Prime Minister Martens and Prime Minister Trudeau. The replies received at the time of going to press are interesting. They indicate clearly that should a nuclear war occur which destroys us all, the political leaders of the world will still be secure, to the bitter end, in the pretence that he or she was not to blame, and that it was all somebody else's fault.

After this, there was an open forum on the prevention of nuclear war. The first speaker was Dorothy Green who said that people are getting tired of the usual twaddle about realism and *realpolitik* and all the rest of it, and that our wisest course might be to fall back on our ancient idealisms. It may well be that the advice in the Sermon on the Mount to love one's enemies contains the wisdom that we all need at this point. We need to touch that spark of humanity which is in all people. Ordinary men and women throughout the world want to be left in peace to get on with the job of feeding and bringing up their families, but those who have a vested interest in war are making this impossible. It is the ordinary, average people in whom we should place our trust. We need to give these people the kind of information that they need, in language they can understand, to enable them to confront those who have plans for their destruction.

Stella Cornelius spoke about the plans of the Peace Programme of the United Nations Association of Australia to organise a Summit for Survival. This would provide a public forum to probe the issues of peace and war, nuclear disarmament, the 'freeze', international security and international reconciliation. The aim would be to create public awareness of the apocalyptic urgency of the international situation, to secure freedom of information for all Australians on the issues, to assist government in the search for solutions, and to engage the media in the vital concerns of survival.

Dr Vladimir Yakubovsky, who had come to Canberra specially for the symposium from the USSR Institute of World Economy and International Relations in Moscow, said that Soviet experts in international relations were pressing their Government to declare a nuclear arms freeze in an effort to break the nuclear deadlock between the USSR and the US. Dr Yakubovsky said that the Soviet Union would agree to an international committee of experts to arbitrate between governments in disputes. In relation to the

Pacific, he pointed out that US missiles are stationed in South Korea and Japan, and that there are plans to deploy US cruise missiles there. He warned that all countries in the Pacific could be drawn into a nuclear war, and said that this risk would be reduced if a peace agreement was signed between all countries of the Pacific. He urged the development of exchanges in economic, scientific and cultural fields between the Soviet Union and other countries.

Greg Weir, Research Assistant to Senator George Georges, drew attention to three proposals from the Queensland Rally for Peace Committee. These were to petition for the setting up of a Senate Select Committee to investigate the establishment of a Ministry of Peace and Nuclear Disarmament, to hold another National Rally for Nuclear Disarmament on Sunday, 15 April 1984 (Palm Sunday) and to campaign to buy the US bases back by sending a peppercorn to Parliament. (The nominal rent which is paid by the US Government for use of the American bases in Australia is one peppercorn.)

Mr Barry Reid, a Member of the House of Assembly of the Australian Capital Territory, asked for suport for the proposal to declare the Australian Capital Territory a nuclear-free zone. Women at the conference discussed plans to surround the American military installation at Pine Gap, and Mrs Catherine Bradfield described how she was distributing the book, *The Hundredth Monkey,* by Ken Keynes Jnr, to inform people about the dangers of nuclear war. Maureen Cummuskey, Mary Leggett, Judith Gates and Kate Blattman presented an excellent musical satire on nuclear war entitled 'Nuclear Images'.

Many other participants were still keen to contribute to the debate when the conference ended. One of them, Bobi Lee Myer, made the important point later that individuals who face squarely up to the threat of nuclear war, and discuss its implications with others who share their concern, feel depressed initially but later feel much better for it. The depression is followed by a strengthening of resolve and a firm determination to do all in their power to prevent the ultimate obscenity. The above are examples of how people from different walks of life can easily devise their own strategies for preventing nuclear war.

It is extraordinarily rewarding for those who are working actively for nuclear disarmament to discover that at long last the tide is turning rapidly in our favour, and that there is an enormous groundswell of support coming from people all over the world. We must all join together, with a sense of urgency, in the struggle for survival.

Michael Denborough

PART ONE: THE NUCLEAR ARMS RACE

1 WILL THERE BE A NUCLEAR WAR?
Frank Barnaby

Frank Barnaby

Few would question that a nuclear world war is the greatest single threat to our society, if not to humankind. Many believe that the probability of a nuclear world war is steadily increasing. And that a, if not the major reason for this increasing probability is that the nuclear arms race between the USA and the USSR is out of political control. Only if, and when, this arms race is brought under control, these people say, will the danger of a nuclear holocaust begin to recede. It is, therefore, crucially important to understand the nature of the superpower nuclear arms race and to try to fathom the causes of it.

The number of ways in which a global nuclear war might break out is frighteningly large. It could start with the deliberate decision of one superpower to attack the other, or by the escalation of a conventional war, or through mechanical error or malfunction in a nuclear weapon system, because of irrational behaviour by those controlling the nuclear alert and firing systems, by the acquisition and use of nuclear weapons by irresponsible governments and, last but by no means least, by the use of nuclear explosives by terrorists.

While the danger of a nuclear war starting by accident, miscalculation or madness is ever present, and, many believe, increasing all the time, the escalation of a regional conflict fought with conventional weapons to an all-out nuclear war between the superpowers is generally thought to be the most likely — more likely than a direct attack by one superpower on the other. This may well be true for today and tomorrow. But will it remain true into the 1990s? or will the danger of a nuclear attack out of the blue, a first-strike in the jargon of the experts, come to dominate? And if one side does get the capability to make an effective first-strike, or even the perception that it can do so, would it be likely to resist the temptation to attack? The answers to these and similar questions are related to the nature of nuclear weapons and the qualitative changes being made to these weapons.

The nuclear weapons now in the arsenals of the nuclear-weapon powers have a vast range of explosive power — varying between the equivalent of that of 10 tons of TNT and at least 20 million tons (20 megatons). It is hard to appreciate what destructive power a 20 megaton nuclear warhead has. It might help to know that the total weight of all the explosives used by man in war throughout history is roughly 20 megatons. Yes, the superpowers really do have nuclear warheads each of which contains as much explosive power as *all the explosives used in war so far put together.*

The nuclear arsenals

Nuclear warheads are of two types — strategic and tactical. Range is the main distinguishing factor between them, the former having very long (intercontinental) ranges, greater than say 6000 kilometres. But the existence of intermediate range missiles, like SS-20s and cruise missiles, confuses the distinction between different types of weapons. The strategic-tactical distinction is, in any case, artificial. Strategic nuclear weapons are deployed on intercontinental ballistic missiles (ICBMs), submarine-launched ballistic missiles (SLBMs), and strategic bombers. Soviet and American ICBMs have ranges of about 11 000 kilometres, modern SLBMs have ranges of about 7000 kilometres, and strategic bombers have ranges of about 12 000 kilometres. Some ballistic missiles carry many warheads — up to 14. Modern multiple warheads are independently targetable on targets hundreds of kilometres apart. These are called multiple independently-targetable re-entry vehicles, or MIRVs in shorthand. Strategic bombers carry free-fall nuclear bombs and air-to-ground missiles armed with nuclear warheads. The most modern of these missiles is the American air-launched cruise missile (ALCM) carried by the B-52 strategic bomber; the ALCM has a range of about 2500 kilometres.

The USA has (end 1982) 1596 strategic ballistic missiles — 1052 ICBMs and 544 SLBMs (SIPRI, 1983). Of these, 1094 (544 SLBMs and 550 ICBMs) are fitted with MIRVs. Three hundred and sixteen B-52s are operational as strategic bombers, carrying 1264 nuclear free-fall bombs, 1114 short-range attack missiles with nuclear warheads and 192 ALCMs. These American strategic nuclear forces carry about 9400 nuclear warheads — 2100 on ICBMs, 4700 on SLBMs and 2600 on bombers. These warheads can deliver a total explosive power of 3500 megatons (Mt): 1500 by ICBMs, 300 by SLBMs and 1700 by bombers.

The USSR has (end 1982) 2335 strategic ballistic missiles — 1398 ICBMs and 937 SLBMs. Of these, 930 (260 SLBMs and 670 ICBMs) are thought to be fitted with MIRVs. Some 150 Soviet long-range bombers may be assigned an intercontinental strategic role. These Soviet strategic nuclear forces carry about 8500 warheads on ballistic missiles — about 5700 on ICBMs and 2800 on SLBMs. These warheads can deliver about 6400 Mt, about 5500 by ICBM and 900 by SLBM. The bombers may carry about 300 nuclear weapons — about a half as free-fall bombs and the rest as short-range attack missiles. The explosive power of these aircraft-delivered warheads totals about 300 Mt.

Tactical nuclear weapons are deployed in a wide variety of systems — including artillery shells, ground-to-ground ballistic missiles, free-fall bombs, air-to-ground missiles, anti-aircraft missiles, atomic demolition munitions (land-mines), ground-, air-, and submarine-launched cruise missiles, torpedoes, naval mines, depth charges, and anti-submarine rockets. Land-based tactical systems have ranges varying from about twelve kilometres or less (artillery shells) to a few thousand kilometres (intermediate range ballistic missiles). The explosive power of tactical nuclear warheads varies from about ten tons to about one megaton. The USA deploys tactical nuclear weapons in Western Europe, Asia, and the United States, and with the Atlantic and Pacific fleets. The USSR deploys its tactical nuclear weapons in Eastern Europe.

We know a great deal about the American nuclear arsenal from official and semi-official sources. We know very little about the Soviet nuclear arsenal from official sources. It is, in fact, extraordinary how much information is publicly available in the USA about American military affairs. The Soviets, on the other hand, are obsessively secretive about military matters. This is not only a Communist habit, it has been true of the Russians for hundreds of years. Sadly, the defence establishments of the other established nuclear-weapons powers, (the UK, France, and China) like secrecy almost as much as the Soviets do. The world would be a safer place if the major powers followed America's example and were more open about their military. Data on nuclear-weapon stockpiles are crucial for an informed debate on nuclear-weapon issues. Much information about the nuclear arsenals of other nuclear-weapon powers has been made publicly available by American intelligence agencies. Available public information about American nuclear stockpiles has been recently collated by the American scientists W. M. Arkin, T. B. Cochran, and M. M. Hoenig.

The US nuclear arsenal has not changed much in quantity and quality for a decade or so. But very considerable changes are planned for the next ten years. The deployment of new weapons is changing policies from nuclear deterrence based on mutual assured destruction, to nuclear-war fighting. Further deployments are likely to stimulate even more far-reaching changes, specifically, to nuclear war-winning policies.

Currently, according to Arkin *et al.* the US nuclear arsenal contains about 26 000 nuclear weapons — down from the peak of some 32 000 reached in 1967. Today's weapons are spread over 25 types — ranging from portable land-mines, weighing a mere 70 kilograms to strategic bombs, weighing about 3.6 tons. As we have seen, the explosive power of the weapons varies considerably — from the equivalent of about 10 tons of TNT for the W54 atomic land-mine to that of 9 million tons of TNT for the B53 strategic bomb. Twelve types of US nuclear weapons are currently deployed in NATO countries.

The numbers of nuclear weapons of different types in the US stockpile varies considerably. The numbers range from 3500 for the W48 155mm nuclear artillery shell to 65 for the W53 Titan-II ICBM warhead. The numbers of nuclear free-fall aircraft bombs deployed total about 7500, spread over five types.

The number of tactical nuclear weapons (about 16 000) is approaching the number of strategic nuclear weapons (roughly 10 000) — a change from the 1950s and 1960s when the US stockpile was mainly tactical. About 6000 tactical nuclear warheads are deployed in NATO.

Six of the 25 types of nuclear weapons in the American arsenal are still in production — the warhead for the air-launched cruise missile, the warhead for the Minuteman ICBM, the warhead for the Trident SLBM, the enhanced radiation warhead (neutron bomb) for the Lance ground-to-ground missile, the 8-inch enhanced radiation artillery shell, and a free-fall bomb.

Four types of nuclear weapons are being, or soon will be, withdrawn from the American arsenal — the Nike-Hercules ground-to-air (anti-aircraft) missile, an atomic land-mine, a strategic bomb, and the Titan-II ICBM. Ten other types will be replaced, and three others partially replaced.

Seven more types of nuclear warheads will be produced over the next five years. These are warheads for the submarine-launched cruise missile, a nuclear ship-to-air missile to defend warships against air attack, a 155mm artillery shell, warheads for the ground-launched cruise missile warheads for the Pershing-II ground-to-ground missile, warheads for the MX ICBM, and a new strategic bomb.

According to current plans, about 9000 new warheads of the six types now in production will be produced. The plans call for the production of a total of about 10 000 warheads for the seven new types to be produced over the next five years. Looking further ahead, to the late 1980s and 1990s, Arkin *et al.* list another seven types of nuclear weapons for production — for anti-submarine warfare weapons, the low-altitude air defence system (an anti-ballistic missile system), the lethal neutralisation system, the corps-support weapon system, advanced tactical air delivered weapons, tactical air-to-surface munitions, and advanced cruise missiles. They estimate that about 10 000 nuclear warheads will be produced for these future weapon systems.

All in all, projected nuclear-warhead production in the USA from now until about the mid-1990s, may involve the production of some 30 000 new warheads, of which about 14 000 are for weapons in current research-and-development programmes. The likelihood is that the US will deploy 23 000 new nuclear warheads by the end of the 1980s. Making an allowance for the fact that about 17 000 warheads will be withdrawn from the stockpile, or replaced during this time, the number of nuclear warheads will grow from the current 26 000 to 32 000 by 1990. The number will, in other words, grow back to the previous all-time record reached in 1967.

The Soviet nuclear arsenal is of about the same size as that of the American one. This means that together the superpowers have deployed about 45 000 nuclear weapons. For comparison, the nuclear arsenals of the other established nuclear-weapon powers (the UK, France, and China) contain a total of about 2000 nuclear warheads. The total explosive power of the American and Soviet nuclear arsenals is roughly 15 000 megatons — equivalent to over one million Hiroshima bombs, or to over 3 tons of TNT to every man, woman, and child on earth, or to 750 times all the high explosive used in all the wars in history. The nuclear arsenals are now, and have for many years been, much larger than needed for any conceivable military, political or strategic reason.

Because of the amount of 'overkill' in the nuclear arsenals, further increases in numbers of nuclear weapons deployed are much less important than qualitative improvements in the weapons themselves. It is the qualitative developments which are causing changes in nuclear policies.

The most important qualitative advances in nuclear weapons are those which improve the accuracy, reliability and targeting flexibility of nuclear weapon systems. Many types of new weapons will be seen as suitable for fighting a nuclear war but not suitable for deterring a nuclear war by mutual assured destruction. Very accurate ballistic missiles can deliver warheads over intercontinental ranges on smaller — and therefore many more — military targets than less accurate ones. In fact, with extremely accurate and reliable nuclear weapons the perception will grow on one side or the other that it will be possible to destroy the enemy's retaliatory capability by striking first.

It is not necessary for one side to possess the ability to destroy completely the other side's retaliatory capability for such a first strike to be contemplated. It is sufficient for the attacker to believe (or perceive) that a surprise attack will reduce the enemy's capability to retaliate to the point where the attacker's casualties, caused by the retaliatory attack, will be 'acceptable' for a given political goal.

In this context one must remember that, in times of crisis, political leaders are more apt to listen to the advice of their military chiefs than to their civilian scientific advisers. The calculations of casualties which affect military decisions are likely to be based on wrong assumptions about the military performance of both sides, the performance of the enemy's weapons are likely to be overestimated and those on our side underestimated; worst case analysis for our performance and best case analysis for their's. Also the calculations about the effects of nuclear weapons are likely to emphasise estimates of prompt deaths and injuries and ignore the uncertain long-term effects, even though these long-term effects may well ultimately be more lethal. In addition, the serious sociological and psychological consequences of the total loss of social and technical services and the trauma of nuclear war are likely to be ignored.

Modernisation of nuclear weapons

The accuracy of a nuclear weapon is normally measured by the circular error probability (CEP), the radius of the circle centred on the target, within which half of a large number of warheads fired at the target will fall. In both the USA and the USSR, the CEPs of ballistic missiles, ICBMs and SLBMs, and of tactical nuclear weapons are being continually improved.

In the USA, for example, improvements have been made in the guidance system of the Minuteman-III ICBM involving better mathematical descriptions of the performance of the inertial platform and accelerometers during flight, and better pre-launch calibration of the gyroscopes. With these improvements, the CEP of the Minuteman-III is about 200 metres, compared with 400 metres for the Minuteman-II. At the same time the design of the Minuteman warhead has been improved so that for the same weight, size, radar cross-section and aerodynamic characteristics, the explosive power of the warhead was increased from 170 000 tons (170 kt) of TNT equivalent to 330 kt.

The new Minuteman-III warheads delivered with the higher accuracy could destroy Soviet ICBMs in their silos (hardened to withstand over-pressures of 1500 pounds per square inch) with a probability of success of about 57 per cent for one shot and about 95 per cent for two shots.

The improved land-based ICBN force significantly increases US nuclear-war fighting capabilities. These will be further increased by the MX missile system now under development. The guidance for the MX will be based on the advanced inertial reference sphere (AIRS), an all-altitude system which can correct for movements of the missile along the ground before it is fired. A CEP of about 100 metres should be achieved with this system. If the MX warhead is provided with terminal guidance, (using a laser or radar system to scan the ground around the target, lock on to a distinctive feature in the area, and guide the warhead with great accuracy onto the target), CEPs of a few tens of metres are feasible.

The MX is a large missile, with a launch weight of about 86 000 kilograms, about 2.4 times heavier than the Minuteman-III, and a throw-weight (the weight the missile can carry as warheads) of about 3500 kilograms. The three MX booster rockets will use advanced solid propellants, very light motor cases, and advanced nozzles to propel the missile twice as efficiently as the Minuteman is propelled. The MX will carry up to twelve warheads, each with a yield of 330 kt. The original plan was to make the MX mobile in an attempt to ensure the survival of many of the MX missiles in a surprise Soviet atack. But the final decision about the basing of the MX has yet to be made.

The most formidable of Soviet ICBMs is the SS-18. This is thought to have a CEP of about 400 or 500 metres, with the accuracy soon being improved to about 250 metres. Each SS-18 warhead probably has an explosive power of about 500 kt. With the higher accuracy a typical SS-18 warhead would have about a 55 per cent chance of destroying a US Minuteman ICBM in its silo. Two warheads fired in succession would have about a 95 per cent chance of success.

The USSR also has the SS-19 ICBM, which is thought to be somewhat more accurate than the SS-18 and equipped with a similar warhead. Some of both the SS-18s and SS-19s are MIRVed, the missiles carrying six, eight or ten warheads. In one of the single-warhead versions, the SS-18 is thought to carry a 20 Mt warhead, which is probably the world's biggest.

The other Soviet MIRVed ICBM, the SS-17, carries four warheads, each with an explosive power of 750 kt. So far, about 670 of the Soviet ICBMs have been MIRVed, compared with 550 US MIRVed ICBMs. The Soviet MIRVed ICBM force carries a total of about 4400 warheads. The US MIRVed ICBM force carries 1650 warheads.

The Soviet strategic ICBM force is an increasing threat to the 1000-strong US Minuteman ICBM force, as the accuracy and reliability of the Soviet warheads are improved. On both sides, the land-based ballistic missile forces provide a nuclear-war fighting element in the nuclear policies, in that the land-based missiles are increasingly targeted on small hardened military targets. The submarine-based strategic forces, however, still provide an element of nuclear deterrence by mutual assured destruction.

Strategic nuclear submarines

The Soviet and American navies operate a total of 95 modern strategic nuclear submarines, equipped with SLBMs. The ballistic missiles carried by submarines are normally targeted on the enemy's cities and industry, and provide the assured destruction on which nuclear deterrence currently depends. A single US strategic nuclear submarine, for example, carries about 200 warheads, enough to destroy every Soviet city with a population of more than 150 000 people. American cities are hostages to Soviet strategic nuclear submarines to the same extent as Soviet cities are to American strategic submarines. Just four strategic submarines on appropriate stations in the oceans could destroy most of the major cities in the Northern Hemisphere.

The most modern operational class of Soviet ballistic missile submarine is the Delta class, carrying twelve or sixteen SLBMs, the missiles having ranges of about 8000 kilometres. The SS-N-18, the missile carried by Delta-class submarines, is the first Soviet SLBM to be MIRVed. These missiles could hit most targets in the USA from Soviet home waters.

The Soviets have (end 1982) deployed 240 SS-N-18s, each equipped with seven MIRVs, each MIRV having a yield of about 200 kt.The SS-N-18s are carried on 15 Delta-class submarines. The other main operational Soviet SLBM is the SS-N-8, with a range of about 8000 kilometres and a single 1 Mt warhead. Two hundred and eighty-nine SS-N-8s are deployed, mainly on 22 Delta-class submarines. The USSR also operates 24 Yankee-class strategic nuclear submarines, each carrying 16 SS-N-6 SLBMs, a 3000 kilometre range missile carrying either a 700 kt warhead or two 350 kt warheads. The two warheads cannot, however, be independently targeted and can only hit targets not very far apart.

In 1980, the USSR launched a very large strategic nuclear submarine, the Typhoon. This 160-metre-long boat displaces, when submerged, 25 000 tons and carries 20 SLBMs. It will become operational in about the mid-1980s and be equipped with a new, more accurate ballistic missile, the SS-NX-20. This SLBM will probably carry twelve warheads over a range of about 8000 kilometres.

By the end of 1982, the USSR had deployed 937 SLBMs, 260 of them MIRVed, in its 62 strategic nuclear submarines. These SLBMs are capable of delivering about 2800 nuclear warheads, about 33 per cent of the total number of warheads in the Soviet nuclear arsenal.

The USA now operates two types of SLBM — the Poseidon and the Trident-I. Each Poseidon carries, on average, nine MIRVs, each with a yield of 40 kt; each Trident-I carries, on average, eight MIRVs, each with a yield of 100 kt.

Trident-I SLBMs are deployed on Trident submarines and on Poseidon submarines. The Tridents are new boats. Two are now in operation. They are approximately twice as large as the Poseidon missile submarines which they are replacing. Each Trident carries 24 Trident SLBMs, with ranges of about 7500 kilometres. Seven more Trident submarines are being built. They should become operational at the rate of about one a year.

Trident-I SLBMs are being fitted into Poseidon strategic nuclear submarines. So far, twelve Poseidons have been fitted with the new ballistic missiles. Nineteen other Poseidon boats are operational; each carrying 16 Poseidon missiles, with ranges of about 4500 kilometres.

The extra range of the Trident SLBMs allows the submarines carrying them to operate in very much larger areas of the oceans and still be within range of targets in the USSR. The submarines do not then have to expose themselves to the same extent to Soviet anti-submarine warfare systems.

The US Navy can deliver 544 SLBMs, carrying a total of about 4600 warheads, with a total explosive power of about 300 Mt. Nearly 50 per cent of American strategic nuclear warheads are carried on submarines.

The Americans plan to increase the accuracy of their SLBMs so that they will become as accurate as land-based ballistic missiles. The CEP of the Poseidon SLBM is about 550 metres; that of the Trident-I is probably about 500 metres at maximum range. The CEPs of SLBMs will be further reduced by the use of mid-course guidance techniques together with a more accurate navigation of the ballistic missile submarines. The deployment of terminal guidance on the warheads will give CEPs of a few tens of metres. SLBMs will then be so accurate as to cease to be only deterrence weapons aimed at enemy cities and become nuclear-war fighting weapons able to destroy enemy strategic nuclear forces.

Soviet SLBMs are significantly less accurate than their American counterparts. Operational Soviet SLBMs are thought to have CEPs exceeding 1000 metres. But the accuracy of the missiles will steadily improve and eventually become as accurate as American ones. The Soviets are said to have already tested a new SLBM, the SS-NX-17, with a CEP of about 500 metres.

Strategic Bombers

The USA is continually modernising its strategic bomber fleet. Currently, B-52s are being provided with air-launched cruise missiles (ALCMs); the plan is to deploy some 3000 cruise missiles, 25 per bomber. The Reagan Administration also plans to build 100 B-1B bombers to replace some of the ageing B-52s. The first B-1Bs should be operational in 1986 and they will also carry ALCMs. The Administration is also encouraging intensive research into the Advanced Technology Bomber, or 'Stealth' aircraft. This programme involves the development of radar-absorbing materials and aircraft shapes to give a very small radar cross-section, as well as terrain-following and other systems to avoid detection by Soviet air-defence systems.

The ALCM is a long-range, sub-sonic, nuclear-armed, winged vehicle, about six metres long, weighing less than 1360 kilograms, with a range of about 2500 kilometres, and a nuclear warhead of about 200 kt. The ALCMs could be launched outside Soviet territory against air defence systems, to destroy their radars and anti-aircraft missiles. Following B-52s would then be able to penetrate into Soviet territory to attack targets with their nuclear bombs and ALCMs. The missiles are accurate enough to be used against small, hardened military targets and, because they have relatively small radar cross-sections are difficult to detect by radar on the ground.

Unlike the USA, the USSR maintains an extensive air defence system based on a family of surface-to-air missiles and a large number of interceptor aircraft. The Soviets will probably extend their air-defence system to be able to cope with the cruise missiles now being deployed by the USA. This will probably involve deploying Airborne Warning and Control System aircraft to patrol constantly the Soviet borders to detect incoming air- or ground-launched cruise missiles, and to alert and control fighter aircraft and surface-to-air missiles to shoot the enemy missiles down. (There are those who believe that the main reason for the deployment by the USA of cruise missiles is to provoke the USSR into spending very large sums on counter-measures.)

Modernisation of tactical nuclear weapons

Both the USA and the USSR are modernising their tactical nuclear arsenals. Tactical nuclear weapons have a shelf-life of 20 or so years after which they must be withdrawn from the arsenal or replaced. This is because the fissile and fusion material in the weapons, particularly tritium, decays and other materials deteriorate. Modernisation is, therefore, inevitable if nuclear weapons are continually deployed.

Among the new types of nuclear weapons planned for NATO are Pershing-II missiles and ground-launched cruise missiles. These weapons are so accurate as to be perceived as nuclear-war fighting weapons; both have a CEP of about 50 metres. Although less accurate than the American missiles, the Soviet SS-20, an intermediate range ballistic missile, is accurate enough, or soon will be made so, to be a nuclear war-fighting weapon, given the relatively large explosive power of its warhead.

The Soviet SS-20, first deployed in 1976, is a two-stage mobile missile, with a range of about 5000 kilometres. About 330 of the missiles are deployed (early 1983); about 60 per cent are targeted on Western Europe (from sites west of the Urals) and the rest on China and possible other Asian targets (from sites east of the Urals). Each SS-20 normally carries three MIRVed warheads, each with a yield said to be about 150 kt. The CEP of the SS-20 is thought to be about 500 metres.

The Pershing-II, to replace the Pershing-I missiles deployed in West Germany, will be provided with a sophisticated new guidance system called RADAG. When the warhead approaches its target a video radar scans the target area and the image is compared with a reference image stored in the warhead's computer before the missile is launched. The computer controls aerodynamic vanes which guide the warhead onto its target with an accuracy unprecedented in a missile with a range of 1800 kilometres. The missile, due for deployment from the end of 1983, will be the only ballistic missile able to penetrate a significant distance into the USSR; it could, for example, reach Moscow from its sites in West

Germany. The plan is to deploy 108 Pershing-IIs with warheads having yields in the low kilotonnage range.

NATO also plans to deploy 464 ground-launched cruise missiles between 1983 and 1988, in West Germany, Italy, the UK, Belgium, and the Netherlands. The missiles will have ranges of about 2500 kilometres and carry warheads with a yield in the kiloton range.

From nuclear deterrence to nuclear war-fighting to nuclear war-winning

Until the end of the 1970s the nuclear policy of the USA was nuclear deterrence by mutual assured destruction. Nuclear deterrence depends on the belief that the enemy will not attack suddenly (pre-emptively) if he knows that most of his cities and industry will be destroyed in retaliation. If the enemy no longer fears that his cities are at risk nuclear deterrence by mutual assured destruction no longer works. This is precisely what happens when accurate and reliable weapons are deployed.

Deterrence is essentially a matter of psychology. What matters is what the enemy believes. It is impossible to maintain a policy of nuclear deterrence by mutual assured destruction with accurate weapons simply because the enemy will assume, willy nilly, that the other side's nuclear warheads are targeted on his military forces and not on his cities. The cities then cease to be the hostages. Accuracy, in other words, kills deterrence. Nuclear-war fighting based on the destruction of hostile military forces, then becomes the only credible, and, therefore, feasible policy.

As we have seen, the current American nuclear policy is a confused mixture of nuclear deterrence by mutual assured destruction and nuclear-war fighting. SLBMs provide the deterrence element, being still targeted on cities, while ICBMs are nuclear-war fighting weapons, targeted on enemy strategic nuclear forces. But when the Trident-II missiles are deployed, about five years from now, SLBMs will also be nuclear-war fighting weapons and the US nuclear policy will become pure nuclear-war fighting.

The worst possible situation would be the deployment of significant numbers of tactical nuclear-war fighting weapons in Europe, together with the deployment of strategic nuclear-war fighting weapons in the USA and the USSR. If tactical nuclear-war fighting weapons are deployed in Europe they will integrate into military tactics at low levels of command. Then, not only would a war in Europe almost inevitably escalate to a nuclear war but the military will more easily come to believe that a nuclear war is fightable and winnable, and that a limited and protracted nuclear war is possible.

Philip Williams (1983) has pointed out that the four main requirements of US strategic policy are as follows: first, to have a capacity to 'render ineffective' the total Soviet (and allied) military and political power structure through attacks on the political and military leadership and associated control facilities, nuclear and conventional military forces and industry critical to military power; second, to possess nuclear forces 'that will maintain through a protracted conflict and afterward the capability to inflict very high levels of damage against the industrial and economic base of the Soviet Union and her allies so that they have a strong incentive to seek conflict termination short of an all-out attack on US cities and economic assets'; third, to 'maintain in reserve under all circumstances, nuclear offensive capabilities so that the US would never emerge from a war without nuclear weapons while still threatened by enemy nuclear forces'; fourth, to improve command and control facilities to a point where they are capable not only of 'supporting controlled nuclear attacks over a protracted period but of maintaining links with those SLBM forces which would be held in reserve throughout the conflict'.

War-fighting deterrence, as it has been called, is giving way to war-winning strategies, in which it is argued that victory is possible in a nuclear war. Typical advocates of the possibility of nuclear victory are Colin Gray and Keith Payne (1982). In a nuclear war, they argue, 'the United States should plan to defeat the Soviet Union and to do so at a cost that would not prohibit US recovery'. They go on, 'a combination of counterforce offensive targeting, civil defence, and ballistic missile and air defence should hold US casualties down to a level compatible with national survival and recovery'.

Over the next five years the US Administration plans to spend about \$25 000 million on strategic nuclear systems, including nuclear defence and command, control, and communications systems. A range of military technologies are being developed which will strengthen nuclear-war fighting and winning perceptions. The most important are those related to anti-submarine warfare, anti-ballistic missiles, and anti-satellite warfare systems.

If one side could severely limit the damage that the other side's strategic nuclear submarines could do in a retaliatory strike, and believed that it could destroy — by, for example, high-energy lasers in space — any enemy missile warheads which survived a surprise attack, then the temptation to make an all-out first strike may well become well-nigh irresistible, particularly during a period of international crisis.

The drift to nuclear world war?

Many believe that the progression from nuclear deterrence by mutual assured destruction to nuclear-war fighting to nuclear-war winning, is considerably increasing the danger of a nuclear world war. Not only is the danger of a deliberate attack by one side on the other increasing as we move towards a first-strike capabilty but, perhaps more seriously, there is an increasing danger of nuclear war by accident, miscalculation and madness. We have seen that this situation is coming about mainly because the nuclear arms race is continually producing qualitative improvements in nuclear weapons and their supporting technologies. Even if political leaders wanted to maintain their old policy of nuclear deterrence by mutual assured destruction they would be prevented from doing so by the characteristics of the new nuclear weapons developed by military scientists. Military science, it could be argued, is no longer under political control.

Today, about 500 000 scientists work only on military research and development, about 25 per cent of all scientists employed on research. This large group of scientists is a powerful political lobby. Moreover, vast bureaucracies have grown up in the great powers to deal with military matters. Academics and bureaucrats join with the military and the weapons industry to form a giant academic-bureaucratic-military-industrial complex intent on maintaining and increasing military budgets and agitating for the use of every conceivable technological advance for military purposes. This complex has so much political power as to be almost politically irresistible. If this is so, the nuclear arms race is now totally out of the control of political leaders. And this is as true in the Soviet Union as it is in the USA.

In fact, because similar military technological developments are taking place in the USSR, Soviet and American nuclear policies must be expected to develop in roughly the same way. In most areas of military technology the USA is ahead of the USSR but the gap is only a few years. The knowledge that the other side will catch up will increase the temptation to make a surprise nuclear attack once the perception of a first-strike capability develops.

Conclusions

I believe that the above evidence leads to the following conclusions:

1. The nuclear arms race is out of political control.
2. Unless the nuclear arms race is soon brought under control there will be a nuclear catastrophe.
3. Nuclear weapons now being developed and deployed are accurate and reliable enough to be seen as suitable for fighting a nuclear war but useless for deterring nuclear war by mutual assured destruction.
4. After another five years, when SLBMs are nuclear-war fighting, superpower nuclear strategies will be pure nuclear-war fighting; nuclear deterrence by mutual assured destruction will be dead.
5. When anti-submarine warfare, anti-ballistic missile, and anti-satellite warfare systems are available, in say 15 years time, the perception of a first-strike will be possible and, therefore, probable; nuclear-war winning strategies may then dominate.
6. The worst situation would be the deployment of nuclear-war fighting weapons in Europe together with nuclear-war strategic weapons in the superpowers.
7. To reduce these dangers it is necessary:
   a) to prevent the modernisation of nuclear weapons in Europe. (This means, in practice, to remove nuclear weapons from the land-mass of Europe, because if nuclear weapons are deployed their modernisation is inevitable.)
   b) To encourage the superpowers to scrap their ICBMs and strategic bombers, which are now so vulnerable as to be obsolete. (This would effectively mean reliance on strategic nuclear submarines and a return to mutual assured destruction.)
   c) The submarines should then be reduced in number,

eventually to zero, as their ballistic missiles become nuclear-war fighting weapons.

8. This programme of nuclear disarmament should be incorporated into a comprehensive disarmament programme, including a comprehensive permanent ban on nuclear-weapons tests, bans on chemical weapons and offensive conventional weapons, control of the global arms trade and the spread of nuclear weapons, and so on.
9. This disarmament programme will only be achieved if the pressure of public opinion is strong enough to overcome the pressure exerted on the political leaders by the military-industrial-academic-bureaucratic complex.
10. To mobilise public opinion it will be necessary to present alternative defence policies to the current policies which rely on the first and early use of nuclear weapons.
11. To avoid the further escalation of the arms race, such an alternative defence policy should be non-nuclear and non-provocative.
12. In a period of low economic growth, the alternative policy should cost no more than current military budgets.
13. Using available and imminent conventional defensive technologies a non-provocative, non-nuclear, militarily-credible, defensive deterrent is possible.
14. The choice for Europe is war or no war because any war in Europe will almost certainly escalate to an all-out strategic nuclear war, whether or not nuclear weapons are deployed in Europe.
15. A non-nuclear, non-provocative defence posture would minimise the probability of a war in Europe.

References

Arkin, W. M., Cochran, T. B., and Hoenig, M. M. (1983), *Nuclear Weapons Data Book*, Ballinger, Washington.

Gray, C. S. and Payne, K. (1982), *Victory is Possible*, Foreign Policy.

Williams, P. (1983), *Deterrence, Warfighting and American Nuclear Strategy*, ADIU Report.

SIPRI Yearbook, (1983), *World Armaments and Disarmament*, Taylor and Francis, London.

2 THE NUCLEAR THREAT TO AUSTRALIA
Desmond Ball and R. H. Mathams

Desmond Ball

Any assessment of the consequences of nuclear war for Australia and its region is plagued by numerous uncertainties. In the first place there are many uncertainties relating to the actual physical effects of nuclear weapon detonations. The devastation at Hiroshima and Nagasaki towards the end of World War II is the only experience the world has had of the results of attack by nuclear weapons. Examination of the consequences to people and property of those attacks and the study of data accumulated from a great many weapon-effects tests have established detailed assessments of the degree of damage that may result from a given nuclear explosion. Even with respect to tests, however, as the classic text on the effects of nuclear weapons emphasises:

> there are inherent difficulties in making exact measurements of weapons effects. The results are often dependent on circumstances which are difficult, if not impossible to control even in a test, and certainly cannot be predicted in the event of an attack (Glasstone and Dolan 1977).

Moreover, the scaling up of test data in order to derive an estimate of the destruction that would result from the multiple detonations that would occur in any large-scale nuclear exchange is not a simple exercise in arithmetic. No methodology exists for assessing the synergistic effects of multiple detonations, although these would undoubtedly be very substantial.

Second, there are uncertainties with respect to the possible scale and targeting of any nuclear attack against Australia. What are the likely targets, how many warheads is any notional adversary likely to allocate to these targets, what size are these warheads likely to be, and are they likely to be detonated in the air or on the surface?

And, third, the consequences of any nuclear attack are dependent upon a number of imponderable factors that are in turn dependent upon the circumstances prevailing at the time of the attack — such as the meteorological conditions, the length of the warning period preceding the attack, the availability and effectiveness of protective measures, the tactics of the attacker, the capabilities for post-attack recuperation and recovery, etc.

It is understandable that, in these circumstances, the occasional public discussions of nuclear threat to Australia should lack direction and tend to stress worst-case possibilities. There has been little informed debate about the likely circumstances in which Australia may be subjected to nuclear attack or the probable scale and location of such attacks, should they ever occur.

Under what circumstances might Australia be subjected to direct nuclear attack?

There are five nuclear weapon powers: the United States, the United Kingdom, the USSR, France, and China. Although one other nation, India, has demonstrated a capability to produce nuclear explosives and a number of other nations have the potential to develop nuclear explosives, international measures to prevent further proliferation of nuclear weapons among nations have so far been effective and may well continue to be so.

All five nuclear weapon powers have the capability to deliver nuclear weapons onto the Australian mainland. Three of them — the US, the UK and France — are friendly to Australia and it is inconceivable that any of them would use armed attack of any kind against Australia. China's intercontinental nuclear capacity is extremely limited (probably less than ten warheads) and in present and prospective strategic circumstances would almost certainly be committed to targets in USSR. The USSR remains as the only nuclear weapon power which realistically could be considered as likely to use nuclear weapons against Australia.

There is a widespread acceptance among strategic analysts that nuclear attack against Australia would occur only in the event of major nuclear conflict between the US and the USSR. The concept of an Australian city or defence-related facility being used as a nuclear 'hostage' by the USSR to bring pressure upon US or Australian decision-makers lacks credibility in circumstances where the US has the demonstrated capability to use tit-for-tat tactics. Moreover, such a ploy by the USSR would destroy its present ability to gain political advantage as an ostensible supporter of world-wide anti-nuclear movements.

The likelihood of major nuclear conflict between the US and the USSR is low. Certainly the two superpowers are adversaries and there is a finite chance that some difference between them affecting the vital strategic interests of one or the other may escalate to nuclear conflict. But there is also much in their strategic relationship to promote stability or at least to reduce the chance of major conflict between them. These positive factors include the essential equivalence of US and Soviet nuclear strategic forces; the capacity and the determination of US and NATO forces to deter attack by Soviet and Warsaw Pact forces; strategic arms limitation negotiations; the attitude of China; a measure of economic interdependence between the two superpowers; and, most important, the knowledge that each has of the other's capabilities. Overall, there is a clear comprehension on both sides of the awesome consequences of escalation to major nuclear conflict. The prospect of large-scale urban/industrial destruction is a powerful deterrent to any use of nuclear weapons.

However, even though the likelihood of major nuclear conflict between the superpowers is low, it is none the less finite. It is therefore both realistic and prudent to consider where and how Australia may be attacked in such an eventuality.

It is not possible to determine precisely what importance the USSR would assign to probable targets in Australia, but some logical assumptions can be made. First, it is clear that the Soviet Union's primary concern would be to reduce, if not destroy, the nuclear forces of the US and its allies and, perhaps, those of China.

Second, the rank order of major US defence and industrial facilities as compared to military forces in Europe and China may be arguable, but all would be ahead of defence and industrial facilities in Australia. Third, although the USSR has deployed more than 7400 nuclear warheads, the actual number available for use against Australian and other subsidiary targets would be quite low indeed. The allocation of Soviet warheads to higher priority strategic targets, the attrition caused by enemy action, the losses due to system malfunctions, and the requirements for a strategic reserve could easily account for the entire Soviet strategic nuclear arsenal. In fact, in a non-generated situation, where the Soviet forces are at their normal day-to-day levels of readiness, the Soviet arsenal would be insufficient to cover the US and NATO conventional military forces, economic-industrial (E/I) and war-supporting capabilities, and administrative and governmental centres.

There are three conceivable categories of nuclear targets in Australia. In descending order of probability, these are the joint Australian-US facilities; Australian defence establishments; and industrial complexes and urban centres.

The nuclear risk to the joint Australian-US facilities was addressed in considerable depth by the Joint Parliamentary Committee on Foreign Affairs and Defence in 1981. In testimony to that Committee, one of the present authors (Dr Ball) argued that 'North West Cape is presently one of the most important links in the US global defence communications network' (Joint Parliamentary Committee on Foreign Affairs and Defence, 1981).

According to official brochures, the base 'may serve several purposes. However, its main reason for existence is to maintain reliable communciations with submarines of the US fleet serving in this area of the world [i.e., the Indian and Western Pacific Oceans]' and, in particular, 'to provide communication for the US Navy's most powerful deterrent force — the nuclear powered ballistic missile submarine' (Holt). The VLF facility for communicating with the American submarines is the largest and most powerful of the two principal VLF stations in the US world-wide submarine communciation system — the other one being at Cutler in Maine (US General Accounting Office, 1979). Pine Gap and Nurrungar are ground stations for satellite surveillance systems — the first concerned with gathering intelligence by the interception of a wide range of electronic signals, including missile telemetry, radar emissions and communications; and the second concerned with the detection of Soviet missile launches and the provision of early-warning to the US National Command Authorities (NCA) (Ball, 1981).

The other author (Mr Mathams) testified to the Joint Committee that

> . . . my view would be that the Soviet Union would certainly consider the North West Cape installation (which obviously is a communications facility and involved with the command of submarines) and probably the other two facilities to be in some way connected with American strategic nuclear forces. As a result, they would feature on the Soviet target list. But I have never been able to assess if they would be near the top or bottom of the list. One could argue quite cogently for either depending on what one believes to be Soviet Perceptions.
>
> However, let us accept that there is a finite risk of their being attacked (Mathams, 1981).

On the basis of this testimony, the Joint Committee concluded that

> it would seem . . . that the Soviet Union is likely to think that the North West Cape installation (which obviously is a facility for communicating with submarines), Pine Gap and Nurrungar are connected in one of the following ways with the American strategic forces:
>
> (a) communications (in the case of North West Cape, including fire orders to ballistic missile submarines);
> (b) early warning;
> (c) target information; and
> (d) signals intelligence.
>
> It can be argued that at least the first two functions mentioned above mean that the facilities would be first order targets in a general war . . .
>
> It would be prudent for Australian defence planners to assume that the joint facilities at North West Cape, Pine Gap or Nurrungar are on the Soviet target list and might be attacked in the course of a nuclear conflict between the two superpowers. In other words, there is a finite risk that one or all of the facilities would be attacked during a Soviet-United States war that in-

SOVIET STRATEGIC NUCLEAR FORCES, July 1982

	Number of Delivery Vehicles	Throwweight Thousand lbs.	Number of Warheads (n)	Yield per Warhead (Mt)	CEP (feet)	Total number of warheads (N)
ICBMs:						
SS-11 Sego	570	2	1	1.0	5000	570
SS-13 Savage	60	1	1	.75	6560	60
SS-17 Mod. 1	125	7	4	.75	1460	500
SS-17 Mod. 2	25	7	1	3.6	1400	25
SS-18 Mod. 1	50	16	1	24	1400	50
SS-18 Mod. 2	68	16	8	.9	1400	544
SS-18 Mod. 3	50	16	1	20	1155	50
SS-18 Mod. 4	140	16	10	.5	900	1400
SS-19 Mod. 2	100	8	1	4.3	1275	100
SS-19 Mod. 3	210	8	6	.55	850	1260
SLBMs:						
SS-N-6 Sawfly Mod. 3	464	1.6	2	.35	4500	928
SS-N-8	298	1.8	1	.8	3000	298
SS-N17	12	2	1	.75	1500	12
SS-N-18 Mod. 1)		2.5	3	.5	2000	
SS-N-18 Mod. 2)	176	2.5	1	2	2000	848
SS-N-18 Mod. 3)		2.5	7	.2	2000	
Bombers:						
Tu-95 Bear	113	8	4	1	3000	452
Mya-4 Bison	43	8	4	1	3000	172
Tu-22M Backfire	75	4	2	1	3000	150
	2579					7419

volved their nuclear strategic forces (Joint Parliamentary Committee on Foreign Affairs and Defence, 1981).

The only other Australian defence establishments to warrant consideration as possible Soviet targets are the naval base at Cockburn Sound (WA) and the RAAF base at Darwin (NT). Both these establishments are under full Australian control, but both are visited periodically by units of nuclear-related US forces. These visits may invite Soviet interest, although the priority the USSR may give to the establishments as nuclear targets would depend on

ALLOCATION OF SOVIET RISOP WARHEADS TO TARGET CATEGORIES IN GENERATED AND NON-GENERATED SITUATIONS, JULY 1982 (RISOP-5G):

	Generated	Non-Generated
Baseline force	7419	7419
Weapons deliverable to target	4840	3352
US SIOP forces	2302	2302
French and British strategic nuclear forces	120	120
Theatre nuclear forces capable of hitting the Soviet Union	250	250
Strategic C^3I targets	400	400
US/NATO conventional/power projection forces	200	50
US/NATO economic/industrial (E/I), war supporting and economic recovery targets	1250	150
US/NATO administrative/ governmental targets	150	50
Reserve warheads	168	30

its assessment of their contributions in the Indian Ocean. At present, US use of both establishments seems to be based more on convenience than necessity.

It has been argued that Australian urban/industrial areas may also be attacked. One such argument relates the possibility of attack to the presence of the joint US/Australian facilities and suggests that the USSR might issue an ultimatum to the Australian Government to dismantle those facilities or suffer nuclear attack on an Australian city. This seems to be an unlikely scenario in all circumstances other than that of extreme tension between the USSR and US. And in that case it could be argued that such a move by the USSR would unnecessarily declare its intentions to use nuclear weapons. Another argument is that the USSR may seek to destroy any Australian capacity to support or succour the US in the aftermath of a nuclear attack. However, following large-scale attacks by the superpowers against each other's urban-industrial areas, their nuclear arsenals would be much depleted. The Soviet Union would probably wish to retain the bulk of its remaining nuclear force as a deterrent against any residual US forces and those of any third country. In any case, there would be many economic recovery targets in Europe and elsewhere far more attractive than Sydney-Newcastle-Wollongong or Melbourne. The likelihood of nuclear attack on Australian urban/industrial areas, even during a major nuclear conflict between the two superpowers, would thus appear to be very low indeed.

In summary, nuclear attack against Australia is conceivable only in the unlikely event of a major nuclear engagement between the US and USSR. Should such an engagement occur, it it likely that the joint US/Australian facilities at North West Cape, Pine Gap and Nurrungar would be attacked with nuclear weapons; nuclear attacks against Cockburn Sound and Darwin would be less likely and may occur only if appreciable use of their facilities was being made by US naval and air forces; nuclear attacks against Australian urban-industrial centres would seem to be most unlikely.

The probable scale and consequences of nuclear attacks against Australia

The Soviet Union has a number of alternative strategic nuclear delivery vehicles (SNDVs) which could potentially be used to attack targets in Australia. These include the single-warhead SS-11 Sego ICBMs, the MIRVed SS-18 ICBMs, various SLBMs carried by FBM submarines operating out of Vladivostok or Petropavlovsk, or strategic bombers such as the Tu-95 Bear and Tu-22M Backfire if staging facilities were made available by Vietnam. The characteristics of these various SNDVs are described in the accompanying table.

All of the potential targets in Australia are 'soft' — i.e., they would be destroyed by the application of relatively low blast overpressures. The VLF antenna at North West Cape and the radomes at Pine Gap and Nurrungar could probably withstand no more than about five pounds per square inch (psi) of peak blast overpressure and, indeed, perhaps only one or two psi of dynamic overpressures. The facilities at Darwin and Cockburn Sound would be destroyed at 25 psi. In the case of urban-industrial areas, substantial damage would be rendered by one to three psi, and the area receiving blast damage at five psi or above would be essentially destroyed. Within the five psi blast area at Hiroshima, for example,

SOVIET STRATEGIC NUCLEAR DELIVERY VEHICLES RELEVANT TO ATTACKS ON AUSTRALIA:

July 1982

	No. of Delivery Vehicles	Range (N.M.)	Throwweight (Thousand lbs.)	No. of Warheads (n)	Yield per Warhead (Mt)	CEP (feet)
ICBMs:						
SS-11 Sego	570	5,700	2	1	1.0	5,000
SS-18 Mod. 1	50	7,450	16	1	24	1,400
SS-18 Mod. 2	68	6,850	16	8	.9	1,400
SS-18 Mod. 3	50	10,000	16	1	20	1,155
SS-18 Mod. 4	140	6,850	16	10	.5	900
SLBMs:						
SS-N-16 Mod. 1		1,300	1.6	1	.7	3,000
SS-N-6 Mod. 2	464	1,600	1.6	1	.65	3,000
SS-N-6 Mod. 3			1.6	2	.35	4,500
SS-N-8	298	4,300	1.8	1	.8	3,000
SS-N-18 Mod. 1			2.5	3	.5	2,000
SS-N-18 Mod. 2	176	4,050	2.5	1	2	2,000
SS-N-18 Mod. 3			2.5	7	.2	2,000
Bombers:						
Tu-22M Backfire	75	3,075	4	2	1	3,000
Tu-95 Bear	113	4,000	8	4	1	3,000

two-thirds of all buildings were destroyed and casualties were approximately 50 per cent dead and 30 per cent injured (Joint Committee on Defence Production, 1979).

Attacks on targets such as these would not require the relatively high accuracy of the more modern Soviet ICBMs, such as the SS-18s, but could be undertaken almost as effectively with the obsolescent SS-11 ICBMs or some of the Soviet SLBMs, leaving the SS-18s for allocation against hard targets such as underground missile silos. There are 120 SS-11 ICBMs within range of Australia, located in three fields at Drovyanaya, Olovyannaya and Svobodnyy (Ball, 1980) but a disadvantage of a single-warhead missile is that one missile must be allocated to each target; indeed, most planners would allocate two warheads (hence two missiles) to each target to compensate for potential reliability problems. On the other hand, a single SS-18 missile with eight or ten warheads could cover all the interesting targets in Australia with two warheads each. The use of SLBMs is also a possibility. The FBM submarines in the Soviet Pacific Fleet between them carry some 312 SLBMs (176 SS-N-6, 72 SS-N-8 and 64 SS-N-18 missiles), the use of which would have the advantage of reducing the warning time to five to fifteen minutes; however, it is unlikely that Soviet planners would choose to send a submarine with twelve or sixteen missiles (and perhaps 48 to 112 warheads) down to the Southern Hemisphere, which might take those missiles out of range of many interesting targets in the Northern Hemisphere, when no more than ten warheads would be needed to cover the whole Australian target set.

Whatever the delivery vehicle chosen, the maximum damage to the sorts of equipment and buildings at each of the possible targets in Australia would be rendered by detonating the warheads in the air rather than at ground level. For a 1 Mt warhead, the optimum height of burst (HOB) for targets of 5 psi is about 9000 feet; for targets of 25 psi, it is about 3000 feet. At these altitudes, there would be very little fallout as compared to a weapon detonated at ground level, since most of the post-explosion weapons debris would be projected into the stratosphere.

The three most likely targets in Australia — North West Cape, Pine Gap and Nurrungar — are fortunately located in relatively unpopulated areas far from Australia's major urban-industrial areas. The VLF antenna at North West Cape is 20 km and the HF transmitter 5 km from the township of Exmouth (which has a population of just on 3000); the Pine Gap facility is 20 km from Alice Springs (which has a population of over 16 000); and Nurrungar is 10 km from Woomera Village (which has a population of 3000).

In the case of North West Cape, the desired ground zero (DGZ) would likely lie south of Tower Zero, calculated not only to ensure the destruction of the VLF transmitter but also to generate 1 or 2 psi over the HF transmitter. For a 1 Mt weapon, this would generate 1 psi over the township of Exmouth — sufficient to shatter windows and cause occasional failures at the joints of panels in standard house constructions, but not sufficient to cause the collapse of standard residential houses.

In the case of Nurrungar, a 1 Mt weapon detonated over the facility would generate about 2 psi over Woomera Village, sufficient to cause the wooden frames of residential-type buildings to shatter so that the buildings collapse and to cause some local fires.

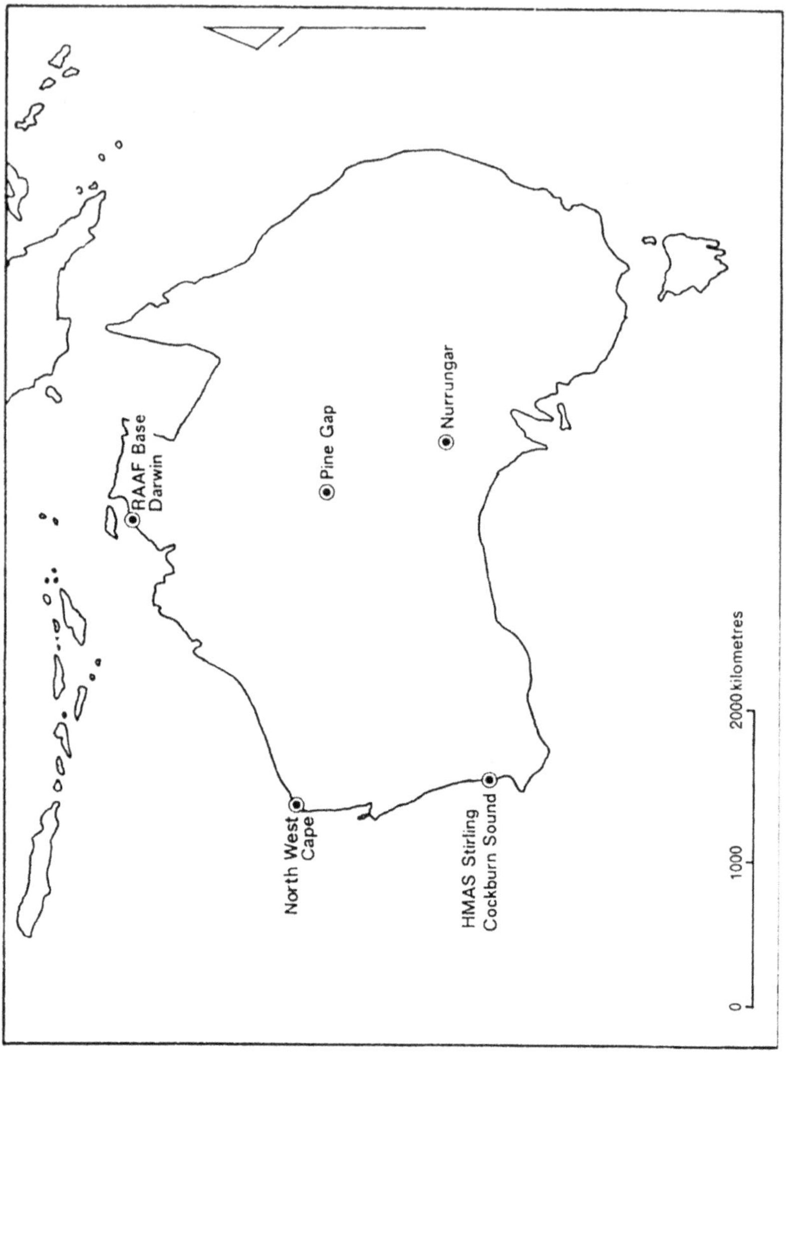
RAAF Base Darwin
Pine Gap
Nurrungar
North West Cape
HMAS Stirling Cockburn Sound
0
1000
2000 kilometres

In the case of Pine Gap, the blast effect at Alice Springs would be less than 1 psi. Some windows would break and there would be occasional frame failures, but most houses would be undamaged. The casualties from blast effects should be nil.

The casualties induced by blast effects from attacks against North West Cape, Pine Gap and Nurrungar are therefore likely to be extremely low indeed — excluding, of course, those employees working at the three facilities at the time, which would total about 500 people. (As at July 1978, there were 657 employees at North West Cape, 474 at Pine Gap and 396 at Nurrungar. For the purpose of this calculation, it is assumed that each of these facilities operates on three eight-hour shifts per day.) However, there could well be casualties resulting from fallout, depending on such factors as the number, size and fission fraction of the weapons used; whether they are detonated in the air or on the ground; the prevailing wind patterns and other meteorological conditions at the time of the attack; and the civil defence measures available (most particularly, the effectiveness of evacuation plans and the nominal protection factors [PF]).

The direction and strength of the prevailing winds vary according to the season and the time of the day (*Yearbook: Australia, 1981*). At North West Cape the winds are generally from the south, so that any fallout generated by an explosion at the VLF and/or HF transmitter sites would blow away from the township of Exmouth. In Central Australia the winds are generally from the south-east, so that fallout from any attack on Pine Gap could well blow over Alice Springs. Around Nurrungar, the winds are much more variable, with northerlies generally prevailing in the winter and southerlies in the summer. Adelaide's winds during the winter are frequently from the north and north-west, so that it could well receive fallout from explosions at Pine Gap or Nurrungar during that season. The worst-case situation for Adelaide would be an attack against Nurrungar which involved a ground burst at a time when the winds were north-westerly and blowing at more than 30 kph (which is quite common in winter), in which case the radiation level in Adelaide would be about 50 to 100 REMs — sufficient to cause nausea and lower resistance to diseases, and to cause some long-term damage, but medical treatment would probably not be required. Under the same conditions, however, (i.e. a 1 Mt ground burst at a time of north-westerly winds), the radiation level over such cities as Port Augusta (population 16 000), Whyalla (32 000),

Port Pirie (15 000) and surrounding areas would be about 300 REMs, which would kill about ten per cent of those exposed, i.e. perhaps more than 10 000 people.

The second category of somewhat less likely targets consists of the bases at Cockburn Sound and Darwin. Although the probability of these being attacked is relatively low (say 20-25 per cent, which in the case of Cockburn Sound is the likelihood of a US nuclear hunter-killer submarine being in port), the consequences of such attacks would be much greater than in the cases of North West Cape, Pine Gap and Nurrungar.

In the case of Cockburn Sound, a 1 Mt weapon detonated in the air above HMAS *Stirling* on Garden Island, WA, would generate blast overpressures of 5 psi out to about 22 500 feet (6.86 km), 3 psi out to 30 000 feet (9.14 km), and 1 psi out to 60 000 feet (18.3 km). The 5 psi contour intersects the mainland only at Cape Peron, which is a non-residential area. The 3 psi contour, within which wooden houses would collapse and some small fires would be ignited, includes much of the residential areas of Peron, Rockingham and Kwinana. The 1 psi contour, within which there would be some slight damage to residential houses, extends up to South Fremantle. Fallout is likely to be a much more serious problem than blast damage, particularly if a ground burst is involved. The afternoon winds over Perth and Fremantle are predominantly south-westerlies throughout the year. A 1 Mt ground burst at a time of such winds would deposit a radiation level of more than 1000 REMs over a narrow but elongated area of Perth and Fremantle, which would kill everyone exposed to that dose. A larger area would receive several hundred REMs, sufficient to kill 50 per cent of the unexposed population. The total fatalities in this worst case situation could well be as many as 100 000 people.

In the case of Darwin, a 1 Mt weapon detonated in the air above the RAAF base would envelope the whole of the built-up area within the 3 psi contour and large areas would receive much higher blast overpressures. The fatalities from blast effects alone could well be as many as half the population (25 000 people), with injuries suffered by a further 40 per cent (20 000 people). Should a ground burst be used, which would more seriously damage the airfield runways while reducing the lethal radii of low blast overpressures against other buildings and equipment at the base, then much of Darwin would be covered with lethal levels of fallout.

The effects of attacks against the third category of possible targets — urban-industrial areas — are the most difficult to assess, precisely because of the relative implausibility of the relevant scenarios. For illustrative purposes, however, assume an attack on Sydney which involved a 1 Mt weapon detonated on the ground at the GPO. In this case, the immediate fatalities (from blast and heat) would number about 180 000, the fatalities from fallout would number about 480 000, and there would be about 350 000 injured people. The great majority of the population would be uninjured (Posenor, 1981). Similar results would attend an equivalent attack on Melbourne.

Warning Times

The consequences of nuclear attack described above have been made on the basis of attack without warning. It is, however, unrealistic to consider that a major nuclear conflict between the US and USSR (which has been postulated as the only circumstance in which a nuclear attack on Australia is likely) would come 'out-of-the-blue'. It is now generally accepted that such conflict would be preceded by a period of international crisis and, probably, a period of conventional conflict. There would be a period of 'strategic warning', of weeks or of months, that would give time for the preparation and perhaps even practice of plans for orderly evacuation of threatened areas for the implementation of measures to reduce damage and casualties and for the procurement of materials for construction of shelters. Such measures could reduce the estimates of casualties given earlier in this paper by a factor of ten or more.

Would Australia be affected, indirectly, by a major nuclear conflict between the US and USSR?

There have been suggestions that, even though Australia may not be subjected to direct nuclear attack in the event of nuclear conflict between the superpowers, it would nevertheless be grievously affected by radioactive fallout from that conflict.

There is, however, no substance to this proposition. The pattern of tropospheric and stratospheric air movement is such that the north and south hemispheres each have a separate circulation.

Most of the debris from nuclear explosions, particularly ground bursts, will fall in the same hemisphere in which the explosions occur. Consequently, the fallout from a nuclear war in which nearly all the detonations take place in the Northern Hemisphere should be deposited in that hemisphere, although some small proportion, projected into the stratosphere from air bursts, may in due time cross the Equator through mixing of the stratospheric air between the two hemispheres (Machta and List, 1960).

The immediate fallout effect on Australia of a major nuclear conflict in the Northern Hemisphere would be negligible. In the longer term, levels of radioactivity in the Southern Hemisphere could increase as a result of transfer of post-explosion weapon debris from the north, but the increase would be gradual and the effects could be moderated by appropriate protective measures (Australia has an effective network of ambient radioactivity monitors). The level of radioactivity in Australia produced by transfer of fallout from the north would probably be only about twice that which resulted from the cumulative effects of atmospheric testing of nuclear weapons in the Northern Hemisphere during the 1950s and 1960s.

Other conceivable effects, such as a diminution of the ozone layer in the stratosphere, in time may affect the entire globe, but current research indicates that these effects would be much less severe in the Southern Hemisphere than in the Northern Hemisphere.

Summary and Conclusions

The probability of direct nuclear attack against Australia is very low. A risk to Australia from attack by nuclear weapons would occur only in the unlikely event of a major nuclear conflict between the US and the USSR. Current developments in military technology and in the politico-strategic relationship between the United States and the Soviet Union are profoundly disturbing, but the prospect of large-scale urban-industrial destruction remains a powerful deterrent to any use of nuclear weapons by the superpowers.

However, although a strategic nuclear exchange is unlikely in the foreseeable future, it must be accepted that there are probable targets in Australia should such an exchange nevertheless occur.

It is not possible precisely to determine the degree of importance that the USSR would assign to these probable targets. It is considered that the Soviet Union has sufficient nuclear warheads deployed to enable it to engage in major conflict with the US and its allies and still retain a capability for attacking a significant number of subsidiary targets on other parts of the globe.

It would be prudent, for planning purposes in Australia, to assume that the USSR considers that the three joint US/Australian facilities at North West Cape, Pine Gap and Nurrungar are necessary to the use by the US of its strategic nuclear forces and that they would be attacked, at some stage, in the event of major nuclear conflict.

The use by US naval and air force units of facilities at Cockburn Sound and Darwin may possibly provoke Soviet attack, but this is considered much less likely than attack against the joint facilities. Nuclear attacks against Australia's urban/industrial areas are only a remote possibility.

The consequences of attack on the three joint US/Australian facilities are not likely to be very great. All three are located away from major concentrations of population and casualties and damage would be limited essentially to the facilities themselves. In the more unlikely event of attack on Darwin or Cockburn Sound, the consequences to areas beyond the RAAF base and naval base would be much more serious; the casualty rate in Darwin would be very high and Fremantle and coastal suburbs would be significantly damaged. In the highly unlikely event of attack against Australia's cities, fatalities could exceed one to two million.

It is generally agreed that there would be a period of strategic warning, of weeks or perhaps months, before nuclear weapons were used by either superpower. This period would provide opportunity in Australia for implementation of civil defence measures that could reduce fatalities by a factor of ten or more.

There is no substance to suggestions that Australia would be grievously affected by radioactive fallout resulting from major nuclear conflict in the Northern Hemisphere.

References

Ball, D. (1981), Testimony in Joint Parliamentary Committee on Foreign affairs and Defence, *Threats to Australia's Security*, pp.16-17.

Ball, D. (1980), 'Soviet ICBM Deployment', *Survival, 22*, pp.167-70.

Glasstone, S. and Dolan, P. J. (1977), *The Effects of Nuclear Weapons*, US

Department of Defence and US Department of Energy, Washington, DC., Third Edition, p.2.
Holt, H. E., *Welcome Aboard the US Naval Communications Station,* (A booklet for personnel newly assigned to North West Cape), pp.10-11.
Joint Committee on Defence Production (1979), *Economic and Social Consequences of Nuclear Attacks on the United States,* US Government Printing Office, Washington, DC, p.29.
Joint Parliamentary Committee on Foreign Affairs and Defence (1981), *Threats to Australia's Security: Their Nature and Probability,* Australian Government Publishing Service, Canberra, pp.16-19.
Machta, L. and List, R. J. (1960), 'The Global Pattern of Fallout', in John M. Fowler (ed.), *Fallout: A Study of Superbombs, Strontium 90 and Survival,* Basic Books, Inc., New York, pp.26-36.
Mathams, R. H. (1981), Testimony in Joint Parliamentary Committee on Foreign Affairs and Defence, *Threats to Australia's Security,* p. 18.
Posener, D. W. (1981), 'Planning for Radiological Defence', *Pacific Defence Reporter, 8,* pp.44-45.
US General Accounting Office (1979), *An Unclassified Version of a Classified Report Entitled 'The Navy's Strategic Communications Systems — Need for Management Attention and Decision-Making',* US General Accounting Office, Washington, DC, p.33.
Year Book Australia 1981, Australian Bureau of Statistics, Canberra, pp.31-43.

3 THE ECONOMIC AND SOCIAL CONSEQUENCES OF PREPARING FOR NUCLEAR WAR

John Langmore

John Langmore

The principal subject of this paper is the economic consequences of military expenditure. Wherever possible the discussion is related to Australia and the region, but the high and growing degree of economic interdependence requires that the subject be discussed globally. The economic and social implications of defence spending are the focus of discussion, not the strategic issues related to the expenditure.

Nuclear forces account for less than 20 per cent of total military spending even in countries with nuclear weapons. For the purposes of this book perhaps only spending on nuclear forces should be discussed, but since nuclear armaments are strategically, bureaucratically and politically integrated with conventional and other forces, and since the important economic and social issues relate to total military expenditure, all military spending is considered. One of the justifications for the development, manufacture and deployment of nuclear weapons used to be their cost effectiveness: they were said to offer 'more bang per buck'.

The paper begins with a brief description of recent trends in military expenditure. In the context of the current international economic disorder the principal economic effects of military expenditure are surveyed: the competition for resources causing reduced consumption and constraints on public and private investment, structural consequences for industry and research and development, the impact on employment and inflation, balance of payments effects, and the consequences for the rate of economic growth. The paper concludes with a few comments about social and political consequences.

Military Expenditure

Military expenditure includes the pay of the armed forces, purchases of hardware and maintenance of existing weapons and other capital stock. There are no agreed or completely reliable comparable figures for military expenditure because there are great technical difficulties in making estimates for individual countries as well as comparisons between them, even when a commitment to impartial accuracy exists, which it often does not (SIPRI, 1982; IISS, 1982; Sivard, 1982; Blackaby and Ohlson, 1982).

For example, the Soviet Union issues just one figure for defence outlays which has stayed the same for a number of years and which excludes important elements such as military research and development (R & D), stockpiling and civil defence. Soviet pricing practices are quite different from those in the West. The CIA compounds this problem by pricing its own estimates of Soviet military production and the armed forces on the same basis as equivalent outlays in the US, producing an entirely illegitimate estimate of an alleged expenditure gap between the Soviet Union and the US.

Even NATO countries differ from each other in their estimates of one another's military spending. The governments of every country try to conceal from potential enemies or more commonly from their own citizens particular areas of military expenditure. A striking example is the initial British expenditures on developing the atomic bomb which was concealed in the civil contingencies fund under the subheading 'public buildings in Great Britain'.

Several independent institutes are widely regarded as making impartial, thorough estimates. Two of these, the Stockholm International Peace Research Institute (SIPRI) and the International

Institute for Strategic Studies (IISS) London, are the sources for most of the following figures. (All figures are in US dollars.)

In 1983 global military expenditure will total around $700 billion. This is more than one twentieth of global income this year, and nearly five times Australia's expected national income in 1983.

About half of world military expenditure is by the superpowers. The US Government estimates outlays on national defence at $215 billion in fiscal 1983, about 6.3 per cent of gross domestic product (GDP). SIPRI estimates the Soviet Union's military expenditure at $115 billion in 1979 which was about 9 per cent of GDP. It is important to remember in this context that military expenditure is not a sound guide to relative military strength because of the wide differences in technological capacity. Other countries with large military expenditures are West Germany ($26 billion, 3.5 per cent of GDP), Britain ($24 billion, 5.4 per cent), France ($24 billion, 4.1 per cent) Saudi Arabia ($24 billion, 20.5 per cent) and Japan ($10 billion, 0.9 per cent). The six largest military spenders account for three-quarters of all military expenditure and almost all military R & D.

World military expenditure has grown rapidly though erratically since the Second World War. There were particularly rapid increases during the Korean and Vietnam wars followed by reductions in the immediately following years. Between 1965 and 1970 real world military spending grew by 5 per cent a year and between 1970 and 1975 by 1.1 per cent. The annual rate of growth between 1975 and 1981 was 2.5 per cent, but a major acceleration began in the US at the end of the seventies.

President Carter initiated a rapid military buildup in the US when he put forward a proposal for an initial 8 per cent real rise in the military budget in the 1981 fiscal year to be followed by 5 per cent annual real increases thereafter. Reagan quickly added to these increases. Real US defence outlays rose by about 10 per cent in fiscal 1982 and real annual increases of about 9 per cent are foreshadowed for each of the following four years. If these plans are realized this will lift US military expenditure from 5.2 per cent of GNP in 1980 to at least 7.2 per cent in 1986.

This US military buildup is not as strong as the growth of defence spending for the Vietnam war. The allocation of national income for military activity in 1986 will be considerably smaller than the peak of 9.4 per cent of GNP in 1968.

Neither SIPRI nor the IISS report any change in the trend annual rate of growth of Soviet military expenditure of about 2 per cent during the last decade.

Australian military spending also rose to a peak during the Vietnam war, of 4.4 per cent of GDP, and declined proportionally in succeeding years. After plateauing for most of the seventies the Fraser Government initiated a major expansion of Australian military expenditure in 1980 and budgeted for a real increase in defence outlays of 5.3 per cent in 1981-82. A five-year defence program (1982-87) was approved in March 1982 which involved annual average increases in real outlays of 6.3 per cent. Mr Fraser planned to lift the proportion of defence spending to 3 per cent of GDP in 1984-85.

An important change in the global pattern of military expenditure during the seventies was the intensification of regional arms races through growth of spending by the better-off developing countries. OPEC members increased their real military expenditure by 17 per cent a year between 1972 and 1981 and middle-income developing countries (GNP per capita greater than $700 a year) by 7 per cent.

China has been reducing military expenditure recently. Indonesian military outlays were also lower in 1981 than a few years before. On the other hand Malaysia, Singapore and Thailand have been increasing their military expenditure in recent years. In 1981 Malaysia spent 8.3 per cent of GNP on its military, Singapore 5.7 per cent, Thailand 3.5 per cent and the Philippines 2.2 per cent. Information on Vietnamese military spending is even harder to obtain than for the Soviet Union.

This brief survey of levels and trends in military spending provides a foundation for examining its economic effects. Another facet of the issue is the economic state of the world.

The Global Economy

The global economy is in the midst of a chronic crisis. The origins of the crisis are complex and have been widely discussed (see for example, Pronk, 1982).

The rate of inflation took off in the late sixties for many reasons. Rapid economic growth in many countries increased inflationary demand-pull pressures. Rapidly rising demand for commodities, to which the Vietnam war and the Russian harvest failure of 1972-73 contributed, led to sharp increases in commodity prices. US deficit financing of the Vietnam war led to balance of payments surpluses in other countries, some of which did not manage their money supply effectively, so enabling faster inflation. The results of the formation of OPEC on oil prices are well known. Money wages began to grow more rapidly in many countries in the late sixties partly to compensate for increasing tax payments as employees moved into higher tax brackets. The collapse of the system of fixed exchange rates added to uncertainty and contributed to growing inflationary expectations.

Most Western governments responded by imposing contractionary policies, and the boom collapsed in 1974 and 1975. Inflation fell but remained much higher than in previous decades and unemployment snowballed, with slower economic growth and rising unemployment in many countries intensifying the recession in others.

A modest recovery began in most of these countries in 1976, but with the OECD urging all members to fight inflation first, and some governments adopting monetarist policies, average growth rates were slower than before 1974. The second oil price shock in 1978-80 depressed real growth rates of oil-importing countries and increased inflation everywhere.

Total real income of Western countries fell in 1982, and this may be the nadir of the current phase of stagflation. Low rates of economic growth (averaging 1.7 per cent p.a. for the OECD area between 1973 and 1980) have resulted in such slow employment growth that over 33 million people (9.25 per cent of the labour force) are now unemployed in Western countries (OECD, 1982). Inflation, which reached an average of 12.9 per cent in 1980 in OECD countries, is now down to about 5.5 per cent, but this is still double the rate of the earlier part of the sixties.

Depressed demand in Western countries has also depressed the export earnings of developing countries, which, in combination with high oil prices, has pushed oil-importing, developing countries into massive current account deficit. This and the tight monetary policies and unprecedentedly high interest rates which many countries in the West have used to suppress inflation has caused severe liquidity problems for many developing countries so that many have had no option but to restrain growth. Yet 750 million people live in absolute poverty, 'a condition of life so characterised by malnutrition, illiteracy, disease, squalid surroundings, high infant mortality, and low life expectancy as to be beneath any reasonable definition of human decency' (World Bank, 1981, p.3 and 1978, iii).

Economic growth rates in the Soviet Union and in other Eastern countries have on average been tending to decline, though they have been more than double those of Western countries during the last three years (World Bank, 1982, p.8).

Clearly the global economy is in disarray. Growing unemployment is increasing poverty in the West and unemployment and under-employment remain chronic problems in the South. The predominant policy prescribes tight money and high interest rates for reducing inflation and this leads inevitably to slow growth and high unemployment. When individual countries such as France and Austria try to break out of this contractionary vice by expanding demand, they are quickly caught by current account deficits which force them back to the policies of restraint and protectionism.

Economic Consequences of Military Expenditure

Competition for Resources

The most obvious economic feature of military expenditure is its opportunity cost, that is, the opportunities which are foregone for alternative consumption and investment. Military expenditure is luxury consumption: every dollar spent on military activity is a dollar which cannot be spent elsewhere. The national accounts make this clear by defining all defence expenditure as public consumption regardless of whether the outlays are on military construction and equipment or on pay or maintenance. Military goods and services have little economic usefulness, either for consumption or for further production. Military expenditure is generally unproductive: it is not aimed at increasing productivity or growth and instead diverts financial and human resources from productive activity.

In the context of a given budget stance military spending is most obviously in conflict with other forms of public consumption. This choice is between defence and such things as education, health, social security or aid to developing countries. Each of these forms of social outlays contributes indirectly to raising productivity as well as to increasing the well being of recipients. President Reagan's cuts in education, health, income security and overseas aid programs to make way for defence are a clear example of the conflict.

The long-term effects are even more serious when public investment is restrained to make way for defence. When trying to reduce outlays governments have less difficulty in reducing spending on capital works than on social security, health and education because they impinge less directly on the electorate. For example, while trying to reduce government expenditure the Fraser Government cut public investment per person by about a quarter, but failed to make a significant impact on current outlays. Yet cancellation or delays in improvements in the public infrastructure reduces national productivity, so slowing economic growth as well as reducing convenience and safety.

The effects are even more generalised when taxes are increased to pay for military expenditure, for this reduces private consumption and investment. The damage to perceived well being may not be significant when the community preference for increased military spending is so strong that there is widened support for an increase in personal income tax, but the long-term effects are more likely to be negative because such an increase tends to reduce aggregate savings and so the flow of investable funds.

Empirical study shows that there is an inverse correlation between the proportion of national income devoted to military expenditure and investment when Western countries are compared (Smith, 1977). Observation supports this conclusion. For example, gross fixed capital formation in the US between 1960 and 1980 averaged 18.2 per cent of GDP, compared with Japan where it averaged 32.5 per cent during the same two decades. Military expenditure in the US fluctuated between five and nine per cent during the same period, while in Japan it remained less than one per cent. The time series evidence suggests that variations in military spending are not strongly correlated with variations in investment within the same country (Smith, 1977; Shultze, 1982, p.32). This may be partly because variations in national aggregate investment tend to be small and to be correlated with the business cycle and interest rates.

Military expenditure may instead be paid for by deficit budgeting, as is happening in the US at present. In fiscal 1983 the US budget deficit is expected to be about $210 billion, almost the same as the outlays on national defence of $215 billion. The deficit is high partly because the economy is depressed, but the increases in defence outlays of around $27.5 billion in each of the last two years have added substantially to total outlays and so to the budget deficit.

This expansion of US defence outlays has been in sharp conflict with the monetary policies of the Federal Reserve Board. Fiscal policy including defence outlays has been expansionary, while monetary policy has been contractionary. The inevitable result has been historically high interest rates which have sharply increased the cost of credit for corporations and home buyers and depressed overall economic activity. The fear of even higher deficits has fueled the inflationary and interest rate expectations of the US capital market so that even when the rate of inflation has fallen, interest rates have remained high. This has continued to depress private investment. The cautious OECD expressed concern about this situation in a review of the US economy last year:

> Given an appropriate budget stance in any given conjunctural (sic) situation, if increased defence expenditure is not to have inflationary secondary effects or crowd out private sector activity through higher interest rates, it needs to be matched by net reductions in other net expenditure (OECD, 1982, p.57).

This summarises well the issue of military competition for financial resources in current circumstances. The damage is not limited to the US however, for historically high US interest rates have forced other countries to raise and maintain high interest rates in order to prevent an outflow of funds and a rapid deterioration in their balance of payments. Thus Australian interest rates have risen in concert with those in the US through the use of monetary policies similar to those in the US in part so as to attract and maintain private capital inflow. The current depression of economic activity in the US caused by these policies has also had an effect around the world, through the reduction in demand for imports. (In earlier times, such as during the Vietnam war, rapid growth of US military spending has led to growth of US imports, which has been of benefit to countries such as Japan and Australia from which those imports have come.)

The current expansion of US military expenditure and the monetarist policies with which it is linked are having a strong impact on Australia. High US interest rates have contributed to high Australian interest rates which have been one of the causes for the sharp contraction in Australian private investment, for the fall in net business profits and for the fall in for housing construction. Depressed US demand has depressed global trade, directly and indirectly contributing to reductions in the volume and price of Australian exports. Stagnation in the US has contributed to stagnation in Australia and so to the growth of unemployment. It would be difficult to attempt to estimate the extent of these effects. The calculations would be extremely complex, for there are both positive and negative effects and many factors other than the high level and growth of military spending are also involved. But it would be difficult to deny that a sequential set of consequences such as that described has occurred.

Military expenditure competes not only for funds but also for personnel, raw materials and energy. Nearly 50 million people are currently employed in defence-related jobs all over the world (UN, 1981, p.88). Although this may create employment (see below) it also involves deprivation of civilian industry and public services of skilled personnel, particularly in specialised occupations. Similarly, the war industry requires substantial inputs of energy and other scarce commodities. A report prepared for the UN on the military use of natural resources shows that there is growing import-dependency in major economies on a dozen strategic raw materials including energy and that cuts in the supply of these could have serious effects on national production (UN, 1981, p.77).

The conflict over the competition for funds and personnel between military expenditure and other uses is particularly sharp in developing countries where needs are even more urgent and funds and skilled personnel are normally in short supply. Yet wars in developing countries have been a feature of the post-Second World War world, and regional arms races are commonplace. The choices for developing country governments must be stark, as they should also be for countries in the north, for military expenditure is competitive with official development assistance. As President Eisenhower said, thirty years ago: 'Every gun that is made, every warship launched, every rocket fired, signifies in a final sense, a theft from those who hunger and are not fed . . .' The tragedy is that neither his own country nor others have heeded his words.

Research and Development

Research and development (R & D) is central to the arms race for it is the quality of technological advance which determines military advantage. So R & D is the field in which the diversion of resources to military purposes is most substantial. Twenty per cent of all highly qualified technical and scientific research personnel are working in military-related research and development projects (UN, 1981, p.80). Military R & D accounts for half of all publicly funded research and development in the US and Britain, more than a third in France but only 2 per cent in Japan.

The opportunity cost of this distortion of research and development priorities is immeasurable because of the damage to the technological dynamism of the civilian sector. The effect is to distort the industrial structure of major weapons-manufacturing countries. The emphasis on military R & D reduces not only the capacity for civilian R & D in major companies but also the strength of the perceived need for civilian research. With civilian research curtailed corporate development becomes concentrated in military production. The contrast of the US and Britain with Japan in the imaginativeness of civilian innovation is readily apparent to consumers in Australia.

Manufacturing industry in those countries has been technically depleted by military R & D. The rate of product improvement in military hardware far exceeds the rate of product improvement in the civilian sector. An estimate of R & D input per unit of output for major Western countries shows that the average military product is about twenty times as research-intensive as the average civilian product (Blackaby and Ohlson, 1982, p.299). Some military apologists have argued that this loss is compensated by the civilian spin-offs which follow from military R & D. This argument is unconvincing because research directed at military goals could not possibly be as cost-effective in achieving particular civilian goals as R & D directed at those goals.

Australia is backward in the proportion of resources allocated to total R & D, but military R & D is a significant component of what little there is. In 1981-82 expenditure on military R & D totalled $145 million. In 1978-79 about a third of Commonwealth government R & D was military, though only a little over ten per cent of total Australian R & D in that year was for military purposes.

Employment and Prices

The effects of military expenditure are, of course, complex: there are economic benefits as well as costs. For example, it is sometimes asserted that military expenditure increases demand and therefore employment and that the Second World War ended the Western depression. Another stream of political theory, Marxism, includes a similar idea, that military expenditure is necessary to maintain consumption and therefore profits in capitalist countries.

Comments such as these can be conceptually misleading for they often involve comparing a situation in which public outlays are increased, in the form of defence expenditure, with a situation in which the increase in outlays does not occur. The depression of the thirties was ended not by the war as such, but by increases in aggregate demand which happened to be in the form of military expenditure. This occurred in a situation in which defence powers also enabled prices to be controlled, which prevented inflation. A threat to national security may provide a strong justification for increased taxation to pay for outlays in defence, but that is a political rather than an economic point. The economic question is whether military expenditure is the most cost-effective means of increasing the rate of economic growth and of employment. A part of the answer has already been suggested: military expenditure tends to divert funds from both public and private investment, so slowing the rate of economic and employment growth in the medium term.

More directly, 'military demand in OECD countries appears to create substantially fewer jobs for a given sum of money than non-military public consumption, but somewhat more jobs, at least at first than private consumption and investment' (ICDS, 1982, p.76). An example of part of this conclusion is given in a detailed study by the US Department of Labor which estimated that a billion dollars spent on defence in 1975 would have created 76 000 jobs compared with 80 000 by local expenditure on health and 104 000 by local expenditure on education. Since then the employment effects of US military spending are likely to have fallen further because of the increasing proportion of that spending which is on high technology weapons. (Other studies with similar conclusions are quoted in the UN Study on the Relationship between Disarmament and Development, 1981, pp.89, 90). Though there do not seem to have been any studies on this type in Australia similar results are likely. Research on the subject would be valuable.

Government outlays also have indirect multiplier effects on employment. In weapons-importing countries such as Australia, the multiplier effect of defence spending is reduced by the import leakage. Fourteen per cent of Australia's total military outlays since 1970 have been made overseas and the proportion is likely to increase following the orders for 75 F/A-18 tactical fighter aircraft and other major weapons.

Military expenditure tends to stimulate inflation. Experience during the Korean and Vietnam wars illustrates this vividly. The Australian consumer price index rose by 22.5 per cent in 1951-52, an unprecedented rate.

A tendency for prices of military equipment to rise faster than consumer prices has been observed in many countries. For example, the US Undersecretary of Defence commented in 1981 that 'costs for defence systems have increased in the last year at a rate higher than [consumer prices] (14.3 per cent) and may approach the 20 per cent level' (Brown and Woolner, 1982, p.160).

There are many ways in which military spending influences the rate of inflation, though the extent to which any of these operate varies greatly from time to time and in different countries. The speed of an increase in defence spending has a powerful influence on the extent of the inflationary effects. For example, the sharp increase in US demand for wool in 1950 for uniforms for soldiers fighting in Korea led to a rapid increase in its price which contributed to inflation in Australia in 1951-52. US stockpiling of commodities has made a noticeable impact on commodity prices on other occasions too.

A rapid increase in military spending may well cause bottlenecks in supply leading to the 'bidding-up' of prices. This is likely to be a particular problem with the current US military buildup. One study has shown that the share of military hardware in the goods-producting portion of GNP (that is GNP less services) is likely to rise from 5.9 per cent in 1981 to 10 per cent in 1986 (Schultz, 1982). Nearly a third of the projected increase in real GNP excluding services is being accounted for by military purchases. The OECD noted this problem in the current US military expansion:

> The planned build-up is sufficiently rapid and concentrated that it could well lead to bottlenecks in capacity, materials and labour skills. If such potential bottlenecks are to be avoided efficiently, the investment repercussions could be considerable; in other words the multiplier effects of the defence programme could be considerable. (OECD, 1982, p.57).

That is, bottlenecks and their inflationary consequences would be avoided or reduced by diverting investment into more productivity-increasing uses, reducing the overall impact on demand and economic growth. Charles Shultz, Chairman of Carter's Council of Economic Advisers has noted the inflationary effects of an abnormally large expansion in the output of a particular group of firms or industries. Costs rise in the war industries as the scramble for materials, components and labour skills causes prices to be bid up. Civilian firms then suffer from the increased costs, and the inflationary consequences become generalised.

Such bottlenecks are one cause of the cost over-runs which have become commonplace in the US. Payments for weapons and military equipment above the original contract price also occur because of inadequate Pentagon supervision, the absence of any effective budget constraint on the Pentagon, telescoping of R & D with manufacture so that planning and testing are inadequate, and lack of the normal competitive disciplines on private firms. The military-industrial firm is not required to give cost minimisation top priority: military forces are concerned more with destructive performance than cost-effectiveness. In fact they have become so pre-occupied with weapons performance that in their thinking performance has become the equivalent of cost-effectiveness.

The military industry is much more concentrated than the economy as a whole. In the US 100 corporations received two thirds of the prime military contract dollars ten years ago (Reich, 1972, p.299) and the proportion is probably higher today. (Amongst those top 100 firms are 20 of the top 25 industrial firms in the US.) Military firms maximise profits by maximising subsidies rather than minimising costs. The Pentagon provides much of the capital and subsidises R & D for military firms. For these reasons profits are higher for military work than for comparable civilian work (Reich, 1972, p.299).

Some economists would also argue that military spending is inflationary because it leads to higher deficits and so faster growth of the money supply. This monetarist analysis is undermined by recent US experience. Between 1980 and 1983 the budget deficit has more than trebled, from $60 billion to an estimated $210 billion, while the rate of inflation has fallen from over 12 per cent to less than 3 per cent. This does not lend support to the analysis in this section either, though the explanation may well be that the severity of the monetary restraint and the resulting national economic deterioration has overwhelmed inflation induced by military expenditure.

Many factors were involved in the continuing Western inflation during the seventies while economic growth was relatively slow. One powerful influence was the struggle for shares of income. Every group in the community aims to maintain and increase the real value of its income: corporations seek higher prices and profits, unions claim higher wages and salaries, trade and professional associations set higher fees, property owners increase rents and the military-industrial complex campaigns for a larger budget. Each group is competing against the others. With prices already rising rapidly, each group has to achieve a regular increase in money income just to keep up. This problem can be addressed indirectly, through contractionary monetary and fiscal policy (which increases unemployment and inequity) or directly, through a prices and incomes policy which aims to restrain all incomes in an equitable way. Restraint of military expenditure would contribute to the success of a comprehensive prices and incomes policy.

Economists have long recognised that the diversion of resources to inefficient industries (through such instruments as protection) retards the economy's potential for economic and employment growth and for reducing inflation. For the reasons discussed the military sector is the most economically inefficient of all, not least because it is cosseted from the normal disciplines of market performance or public surveillance. The diversion of resources to the military sector therefore worsens the trade-off between inflation and unemployment, so that for a given rate of inflation unemployment is higher or for a given rate of unemployment inflation is higher.

Balance of Payments Effects

Trade in weapons intensifies balance of payments problems for importing countries but benefits arms exporters. The annual rate of total arms exports has reached $35 billion (Sivard, 1982) with the USSR (37 per cent), the US (34 per cent), France (10 per cent), Italy (4 per cent) and Britain (4 per cent) being the principal exporters of major weapons (SIPRI, 1982, p.176). Industrialised countries import 38 per cent of major weapons, countries in the Middle East 27 per cent and the remainder are imported by other developing countries. So imports of weapons cause a significant loss of international purchasing power for many countries and the problems this creates are compounded by the fact that military imports generate no income or export with which to service the added international debt. Their capacity to purchase non-military goods and services is also reduced, which effects the trade of countries like Australia which do not have significant military exports.

The attempt to restrain the global arms trade initiated by President Carter broke down. The Conventional Arms Transfer talks between the United States and the Soviet Union were postponed indefinitely four years ago. The drive to increase arms exports is being exacerbated by international tension, by the strategic advantages of being an arms-exporting country and by the economies of scale available when the volume of production increases. The sophistication of the weapons salesmen was evident when Australia was deciding on a new order for tactical fighter aircraft. Arms sales to the Middle East escalated massively following the first oil price rise. The arms trade has been an important factor in the recycling of petrodollars and in so doing has intensified the Middle East conflict. Australia's imports of military equipment and stores have averaged about $300 million during each of the last four years, not a major drain, in fact only about 2 per cent of imports, but a leakage which the country would be better off without. The large orders for major weapons to be delivered during the eighties mean that the import leakage is likely to be considerably higher during the next few years.

The Consequences for Economic Growth

Assessing the impact of military expenditure on economic growth involves summarising the influences already discussed. Since military expenditure tends to reduce public and private investment, to reduce labour productivity, divert funds and personnel from civilian research and development and to retard employment growth, it tends to retard the rate of economic growth. The inflationary consequences and weakened balance of payments for arms importers caused by military expenditure have probably contributed to justifying the contractionary policies applied by some countries, which have also depressed their economic activity and growth. This is not to argue that military expenditure cannot stimulate economic activity but only to emphasise that economic growth would be faster if the resources used for military purposes were instead applied directly to achieving the goals of economic and employment growth.

Studies of industrial countries have shown a high correlation between military expenditure and low rates of economic growth

MILITARY EXPENDITURE, R & D AND ECONOMIC PERFORMANCE

	Defence Expen/GDP %		Defence R & D/GNP %	GDP per capita rate of growth p.a. %	
	1970	1980	1975	1967-73	1973-80
USA	8.0	5.6	0.64	2.5	1.3
UK	4.8	5.1	0.62	3.0	0.9
France	4.2	4.1	0.35	4.7	2.4
Germany	3.3	3.3	0.14	4.5	2.4
Japan	0.8	0.9	negligible	8.0	2.7

Sources: *SIPRI Yearbook, 1982;* R. Huisken, 1982; OECD, *Historical Statistics 1960-1980.*

(UN, 1981, p.81; Sivard, 1977; Huisken, 1982). The economic performance of countries such as Japan and Austria which spent less than 1 per cent and around 1.2 per cent of GNP on military outlays respectively clearly support this conclusion. Conversely, a striking example of the damage to an economy from high military expenditure is Israel which spends 30 per cent of its GNP on defence. Israel has almost destroyed itself in order to save itself.

One controversial econometric study of the impact of military expenditure on economic growth in developing countries concluded that high defence spending actually caused growth rates to rise (Benoit, 1977). Other researchers have criticised this conclusion by showing that problems of data reliability, specification of variables, sample size and type and estimation methods invalidate Benoit's conclusions. Smith and Smith (1980) concluded that there was no econometric evidence for a systematic relationship between military expenditure and economic growth in developing countries and explained this by the diversity of the roles of the military forces in those countries. For example military governments in some countries use repressive methods to extract resources from the population, so perhaps increasing investment above the level which a democratic government could achieve. Such policies can lead to growth without development through increasing inequality and poverty (Mack, 1983).

This discussion has shown that the net negative economic consequences of military expenditure arise principally from its opportunity cost — from the opportunities foregone for alternative use of resources. Military expenditure can also have negative social and political consequences.

Social and Political Consequences

The level of total military expenditure during the last thirty years is unprecedented in peace time and has been associated with the permeation of a culture of militarism through industrial countries. Militarism involves acceptance of the use of military powers as a means of controlling relations between states. Militarism influences institutions such as government and the armed forces as well as the wider community. The reaction of the British people to the conflict over the Falkland Islands was a clear example of the pervasive influence of militarism. The widespread acceptance and endorsement of the arms race is another symptom.

The ideology of militarism has enabled democratic freedoms and rights to be eroded to some extent in most countries, and substantially in many. The powers of secret information-gathering services have been considerable in most Western countries in the post-war period, though the climate of opinion is perhaps less paranoic now than at the height of the cold war. Recent relevations about ASIO however, suggest that this may be due to conplacency rather than to a reduction in their activities. In Eastern countries, human freedoms are much more severely curtailed, and dissidents are sometimes denied human rights. So the political consequences of the spread of militarism can be severe.

A central element in the institutional framework for the spread of militarism has been the growth of the military-industrial complex in weapons-producing countries. For example firms in the US war economy operate differently from other private enterprises, for their operations are commonly controlled from the Pentagon, and they in turn are likely to have strong political influence. Military-industrial firms are not autonomous, the Pentagon retaining final decision making control over the quantity, method of production, price and distribution of the products of these firms. Nor do they minimise cost, in part because of the absence of effective budgetary restraint on military expenditure. These military-industrial firms have all the imaginable worst characteristics of nationalised industries but offer none of the benefits.

There is a high degree of competition between members of the US Congress to influence military expenditure because of the regional implications of the location of military production. The significance of military investment is indicated by the fact that in 1970 for every dollar of corporate after-tax profits there was $1.50

available to the Department of Defence (Melman, 1972, p.315). This encourages a massive lobbying effort because of the value of success to particular corporations, cities or regions, and this inevitably distorts political decision-making and the structure of the economy. It means too that representatives of those regions where military production is located have a strong vested interest in supporting military expenditure.

The influence of militarism in developing countries has been extensive. By the end of the seventies military government was the rule rather than the exception in Latin America, Africa and Asia (Mack, 1983). In many of these countries the strategy adopted for economic growth has increased inequity and sharpened social conflict and the military forces have intervened to keep order, at the cost of increased military expenditure and reduced human rights.

Conclusion

Given the economic, social and political costs of military expenditure, the obvious question which arises is why does it continue. This has been the subject of many papers. Such factors as the momentum of research, the instinct for institutional survival in the armed forces, and the self-interested lobbying of military firms, combined with the real and/or perceived threat of military aggression tend to maintain existing patterns.

Military expenditure is the symbol of political responsibility for national security. Political arguments about commitment to national security are commonly about the proportion of national income allocated to military activity rather than about the more important but complex issues of defence strategy and appropriate armaments. The current US military buildup was not based on any costing of the proposed programs but on the desire to given expression to a commitment to enhance American power. David Stockman, the Director of the Office of Management, is quoted as remarking that 'the defence program . . . was just a bunch of numbers written on a piece of paper' (SIPRI, 1982, p.103).

Is there any hope of cutting military expenditure? The fact that the proportion of national income devoted to military spending has fluctuated in many countries, and fallen significantly in some over considerable periods, gives modest grounds for hope. The pressure on civilian governments to make room for increased outlays on public consumption and investment by restraining defence expenditure is often strong and influential.

Sustained and substantial reductions in military spending would require countries to develop a new formulation of national interests that involved reduced military commitments. For the US that might mean reducing the commitment to defend Western Europe, which cost $81 billion in 1981, half total US military outlays. The debate in Australia about whether to buy an aircraft carrier has in part been of this type. A similar issue for Australia is about the justification for maintaining a squadron in Malaysia. A minimal requirement is greater attention to cost-effectiveness in planning defence through, perhaps, introduction of less costly weapons and reducing the number of defence force personnel.

Sustained reduction of military expenditure would involve disarmament. Multilateral arms negotiations may lead to arms control or even arms reduction but it is plausibly argued that eventually unilateral arms reductions will be required to initiate effective disarmament (Mack, 1983). Most countries could think of steps which would unambiguously reduce the threat to potential enemies without involving loss of adequate defence. For example, the US could dismantle land-based missiles without loss of nuclear deterrence because there are more than enough submarine launched missiles (La Rocque, 1980; Mack, 1983). Australia could refuse to export uranium because of the integration of civilian and military uses in the nuclear fuel cycle, scrutinise much more carefully the purposes and uses of US bases, and refuse to expand defence spending unless there is decisive evidence of the necessity for particular weapons or programs.

These suggestions are notes only: more research and creative thinking is required. Australia urgently needs a Peace and Development Research Institute where impartial scholars and educators would gather information, analyse problems and suggest possibilities for approaching issues such as those discussed in this paper.

We live in a barbaric age. The holocaust and destruction of World War II could be merely an overture to the Armageddon which the superpowers now have the capacity to cause. Their activity is not remote from us. Not only is our survival threatened, but the economic cost of their activity severely damages the current wellbeing of people everywhere. Military spending is reducing our living standards, increasing the rate of inflation and the number of people who are unemployed. Major economic and social problems are exacerbated by the arms race. More arms make humankind poorer, not safer.

References

Blackaby, F. and Ohlson, T. (1982),'Military expenditure and the arms trade: problems of data', *Bulletin of Peace Proposals, 13,* pp.291-308.

Benoit, E. (1977), 'Growth and defence in developing countries' *Economic Development and Cultural Change, 26,* p.2.

Brown, G. and Woolner, D. (1982), *A New Aircraft Carrier for the Royal Australian Navy,* Working Paper No 57, Strategic and Defence Studies Centre, ANU, July.

Calleo, D. P. (1981), 'Inflation and American power', *Foreign Affairs, 59,* pp.781-810.

Carlson, K. M. (1981), 'Trends in federal spending: 1955-86', *Federal Reserve Bank of St Louis Review, 63,* pp.15-24.

Gordon, M. (1981), 'If defence spending is on the rise can inflation be very far behind?', *National Journal,* pp.1101-5.

Huisken, R. (1982), 'Investing for security in the twenty-first century', *Australian Outlook, 36,* p.1.

Independent Commission on Disarmament and Security Issues (ICDSI) (1982), *Common Security: A Programme For Survival.* A Report prepared under the Chairmanship of Olaf Palme, Pan, London.

International Institute for Strategic Studies (1982), *The Military Balance 1982-83,* London.

Kurth, J. R. (1972), 'The political economy of weapons procurement: the follow on imperative', *American Economic Review, 62,* pp.304-11.

La Rocque, G. (1980), 'The defence budget controversy', *Challenge,* May-June, pp.37-43.

Langmore, J. and Peetz, D. (1983), *Wealth, Poverty and Survival,* George Allen and Unwin, Sydney.

Mack, A. (1983), 'Militarism or development: the possibility for survival', in Langmore and Peetz, op cit.

Melman, S. (1972), 'Ten propositions on the war economy', *American Economic Review, 62,* pp.312-8.

—, (1979), 'Inflation and the Pentagon's budget' *America,* 30 June, pp.532-533.

Organisation for Economic Cooperation and Development (1982) *United States,* Paris, June.

—, (1982a), *Economic Outlook,* Paris, December.

—, (1982b), *Historical Statistics 1960-1980,* Paris.

Pronk, J. P. (1983), 'The new international economic order: a second look', in Langmore and Peetz, op cit.

Reich, M. (1972),'Does the US economy require military spending', *American Economic Review, 62,* pp.290-303.

Saltman, J. (1977), 'Economic consequences of disarmament', *Peace Research Reviews, 7,* pp.53-6.

Schultze, C. L. (1982), 'Do more dollars mean better defence?', *Challenge,* January-February, pp.30-35.

Smith D. and Smith R. (1980), 'Military expenditure, resources and development' Dept. Economics Birkbeck College, London.

Smith, R. P. (1977), 'Military expenditure and capitalism', *Cambridge Journal of Economics, 1*, pp.61-76.

Stockholm International Peace Research Institute (1982) *World Armaments and Disarmament Yearbook 1982,* Taylor and Francis, London.

Sivard, R. L. (1982), *World Military and Social Expenditures,* Lasburg, Va.

United Nations (1978), *Economic and Social Consequences of the Arms Race and of Military Expenditures: Updated Report of the Secretary General,* UN, New York.

——, (1981), *Study on the Relationship Between Disarmament and Development: Report of the Secretary-General,* UN, New York, A/36/356.

United States (1978), *Economic Report of the President,* USGPO, Washington.

World Bank (1978), *World Development Report*, OUP, Washington.

——, (1981), *World Development Report,* OUP, Washington.

——, (1982), *World Development Report,* OUP, Washington.

Comments by H. C. Coombs, Di Langmore, David Peetz and Derek Woolner were helpful in revising a draft of this paper but responsibility for the paper is entirely my own. No part of the paper should be interpreted as reflecting policy of the Labor Government.

PART TWO: THE CONSEQUENCES OF NUCLEAR WAR

4 THE MEDICAL CONSEQUENCES OF NUCLEAR WAR

Michael Denborough

Michael Denborough

This book now takes a significant change in direction, as it moves from considering nuclear missiles to examining the effects that these ghastly things have on the lives of women, men and children. Whilst I realise that some people are impressed by the enormous amount of energy released by atomic explosions, the only emotions that the mushroom cloud arouses in me, as a practising doctor, are those of horror and revulsion. I am filled with horror when I think that one-quarter of my fellow research scientists throughout the world are engaged in armaments research, and I have a feeling of total revulsion when I consider the effects of atomic weapons on the lives and well-being of women, men and children. For, let us make no mistake about it, the purpose of nuclear weapons is to exterminate women, men and children.

Physicians throughout the world are uniting to halt the nuclear arms race. If the medical consequences of nuclear war are considered, it is easy to understand why the conservative medical profession has decided to join in the international debate about the increasing danger of nuclear war.

The damage to humans after a nuclear explosion is produced by the effects of blast, heat and radiation. About one-half of the energy generated by the atomic bomb is given off as blast. The front of the blast moves as a shock wave — a wall of high-pressure air, spreading outward at a speed equal to or greater than that of sound. It travels about 11 km in 30 seconds. The shock wave is followed by a hurricane-force wind.

About one-third of the total energy generated by the bomb is given off as heat. The fireballs produced by the nuclear explosions instantly reach temperatures of the same magnitude as that of the sun (several million degrees Celsius). The fireballs grow to their maximal diameter of about 400 metres within a second. At a distance of 500 metres from the centre of the explosion the thermal radiation in the first three seconds is about 600 times as hot as the sun on a bright day, and even at 12 km it is still sufficiently hot to burn human skin.

About 15 per cent of the energy generated by the bomb is given off as ionizing radiation, about a third of which is emitted within a minute of the explosion. The remainder is emitted as fallout from radioactive materials. The radiation dose at the centre of the explosion is lethal and is of the order of 100 000 rads. Those who are further away and receive only 1 per cent of this dose will die quickly, and those receiving 0.5 per cent of the dose will die within a month. Those who receive smaller doses of radiation have a variety of symptoms and signs. Resistance to infection is very much reduced and septicaemia is a common cause of death.

Medical Consequences

Victims of nuclear weapons who die immediately are usually crushed or burnt to death. If a one megaton bomb were to be exploded over a city, 98 per cent of humans within a radius of 4 km would be killed outright. Within a radius of 7 km from the blast, most buildings would be destroyed. Half the people in this area would be killed, and another 40 per cent would suffer terrible multiple injuries. The blast would transform debris from destroyed buildings into missiles which would cannon into human beings, and the people themselves would be transformed into missiles and would be hurled into any immovable objects which were still

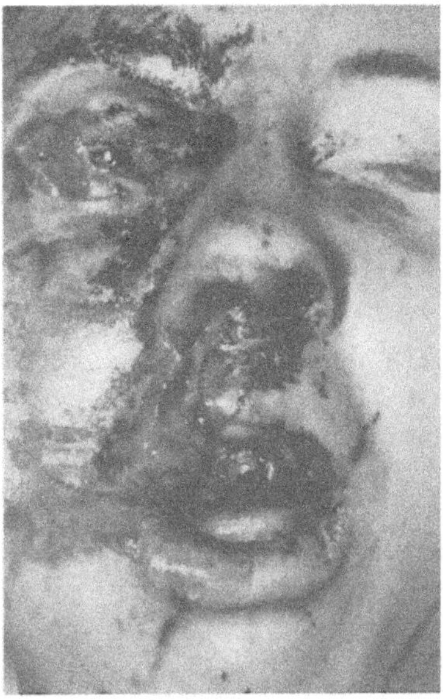

Trauma to the head and face

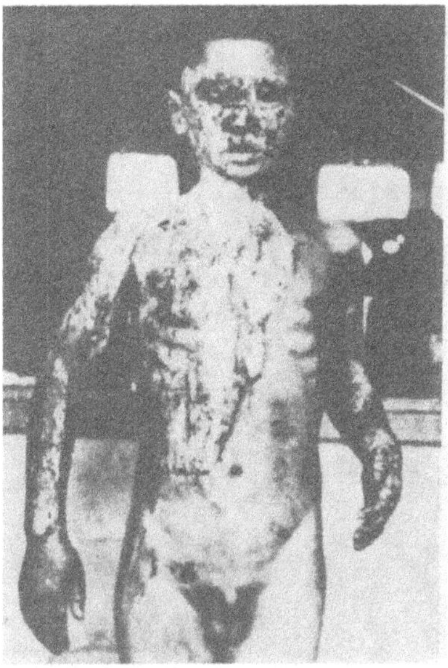

Sixteen year old Nagasaki victim with burns to the face, upper half of body and arms

standing after the explosion. As far as 21 km from the centre of the blast, buildings and humans would still be severely damaged.

It is not possible to list all the injuries which would occur. They include haemorrhage, severe head, chest and abdominal injuries, multiple fractures and lacerations. If close enough to the blast, permanent deafness would result from perforated ear drums, and if the victim was unfortunate enough to have glanced reflexly at the flash of light from the explosion, he or she would be permanently blind.

Burns are particularly important. The thermal energy released by the nuclear explosion produces flash-burn casualties and by causing fires, leads also to flame-burn injuries. Near the centre of the explosion, humans would be incinerated. The radius within which skin burns occurs is related to weather conditions. Thermal radiation, like sunlight, is less intense in areas where it is indirect. Exposed areas of skin are therefore most vulnerable. After a one megaton airburst, full-thickness or third degree burns occur up to a

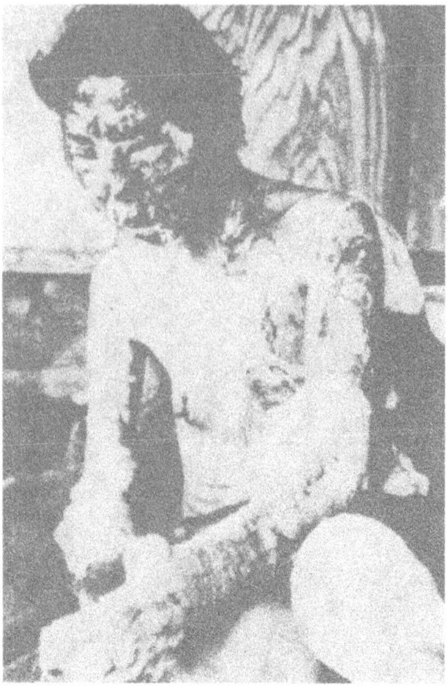

Forty five year old Nagasaki woman victim with burns to the face and upper half of the body. She died on 15 October 1945

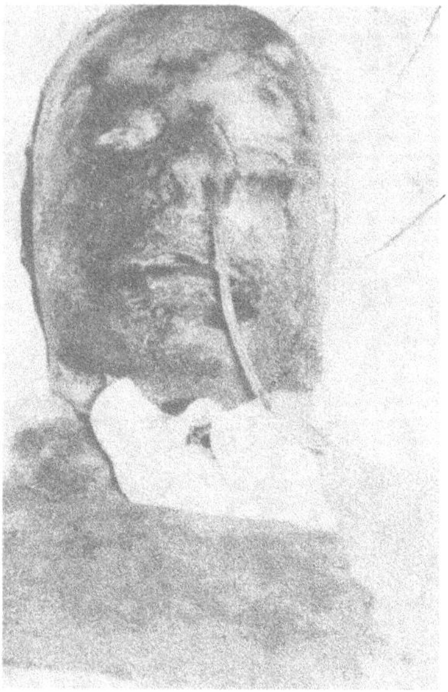

Burns to the face and chest

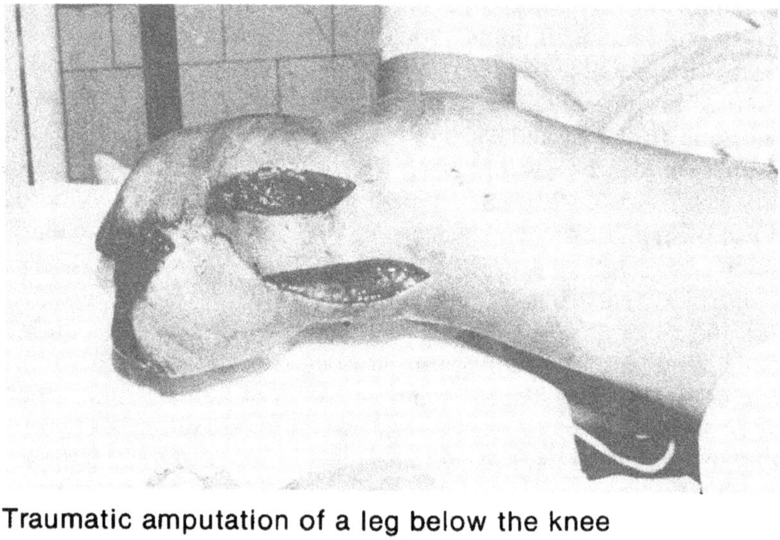

Traumatic amputation of a leg below the knee

distance of 8 km. Second degree burns with blisters which can lead to infection and scarring occur up to a radius of 10 km, and first degree burns, like severe sunburn, occur within a radius of 12 km.

Thermal radiation may also ignite beds and chairs within houses, and after a one megaton blast, would ignite fires within a radius of 12 km. Fuel storage tanks would be ignited starting uncontrollable fires which can exceed a temperature of 1000°C, melting metal and glass. In these areas, oxygen is consumed by the fires in such vast quantities that humans die from asphyxiation. Another serious consequence of fires would be the inhalation of smoke leading to serious damage to the lungs.

Radiation produced by nuclear explosions is of two kinds. Direct radiation is produced at the time of the explosion, whereas delayed radiation or fallout is produced over a longer period. Direct radiation produces 'radiation sickness' which is a complicated condition affecting many organs, and requiring sophisticated investigation and treatment. The initial symptoms of nausea, vomiting and diarrhoea occur in individuals exposed to moderate, sub-lethal or lethal doses of radiation. This means that it is virtually impossible to distinguish those patients with 'radiation sickness' who might benefit from treatment, from those who will die no matter what is done for them.

Radioactive fallout is also a serious threat to human survival, both in the short term and from delayed effects. The amount of fallout differs depending on whether the nuclear weapon explodes at ground level or in the air. An airburst produces less radioactive material than a ground burst. A ground burst activates dirt and debris and lifts these particles high into the atmosphere. The heavier particles fall rapidly over a small area, reaching areas where people have already been killed or severely wounded by blast and thermal injuries, making rescue attempts hazardous. The lighter particles which rise higher are carried by winds before falling back to earth. The amount of radiation received from fallout by any one individual is unpredictable, and depends on weather conditions. Someone 15 km away from the nuclear explosion may receive a very small exposure to radiation, whereas someone 150 kms away may receive a lethal dose.

Prominent long-term effects of radiation damage from nuclear explosions are leukaemia, the incidence of which is increased about 30 times, and other malignant tumours. Children born to women who are pregnant at the time of the radiation damage show an increase in congenital defects (particularly microcephaly with mental retardation).

Radioactive fallout contaminates a huge area beyond the area of destruction, and makes it uninhabitable for up to a year afterwards. The main hazard from early fallout (within 24 hours of the explosion) is from iodine^{131} which enters humans from the consumption of milk from cows grazing on contaminated land. Contaminated milk can be detected thousands of kilometres from the explosion sites. The main radioactive hazard from long-delayed fallout is from strontium^{90} and caesium^{37}. Their half-lives are so long (28 and 30 years) that the delay in deposition decreases their activity very little. Strontium^{90} reaches humans mainly through milk and meat, while caesium^{37} enters through fish, vegetables and other plants.

Nuclear War in the Northern Hemisphere

In a recent volume of *Ambio,* an environmental journal published by the Royal Swedish Academy of Sciences, an examination is made of the medical consequences of a nuclear war between the USSR and the US in which less than half of the total explosive power in the Soviet and American nuclear arsenals will be used (Middleton, 1982). It is estimated that out of a total urban population of nearly 1.3 billion people, one half would be either crushed or burnt to death. One-quarter would be seriously injured.

In case it is thought that these figures are exaggerated it is worth noting the recent report produced by five eminent doctors for the British Medical Association (1983) after 20 months study. They considered that an attack by the Soviet nuclear weapons currently targeted on Britain would kill 38.6 million people and injure another 4.3 million. A one megaton airburst over St Paul's Cathedral in London would result in 1.6 million blast injuries and up to 650 000 severe burns. They concluded that the burden of just one bomb, dropped on a city, would completely overwhelm the medical facilities of the United Kingdom.

The survivors from a nuclear attack would be faced with enormous medical problems. Decaying corpses would litter the streets, and millions of initial survivors, burnt, irradiated and starving, would die slowly and in agony. There would be few medical personnel to help them, and no drugs or equipment to ease their death. Delay in treatment would result in a high incidence of wound infection. Ruptured drainage and sewage systems, together with the presence of decaying corpses and animal carcasses would increase enormously the hazards of infection. Radiation sickness would add to the problems of those already wounded. Also, the destruction of food, water safe for drinking, shelter, fuel and power supplies would assist the spread of disease. Food stockpiles within 10 to 20 km of the nuclear explosion could be exposed to fallout. Farm animals are sensitive to radiation, but insects and vermin are much more resistant. The proliferation of flies, mosquitoes, cockroaches and rats would lead to epidemics of typhoid, cholera, typhus and malaria. Children would be susceptible to epidemics of polio, diphtheria and tuberculosis.

Psychological Consequences

So far only the physical consequences of nuclear war have been described. The psychological effects of a disaster of this magnitude can only be conjectured. Profound effects on behaviour would be expected. An act of such gross inhumanity would lead to grave anxiety and depression, and many would doubt the value of continuing their existence. Aggravating these initial reactions would be fear about the effect on their health of radiation and infectious disease.

Opposition to Nuclear War

In the face of this grim picture, it is surprising that the nuclear arms race has not generated more opposition throughout the world. A careful psychological analysis of this paradox has been made by Nicholas Humphrey (1981). He describes powerful inhibitory forces which block or deflect us from the grave danger that we face. These include incomprehension and denial, social embarrassment, helplessness and the Strangelove syndrome (latent feelings of admiration, almost of appetite, for 'the Bomb' and the final solution that it provides).

Humphrey says that in the US opponents of nuclear war may be silenced by the law, in the USSR they may be locked up in a mental hospital. In Britain and Australia the technique of social pillory may be used. Anyone who forces an unwanted confrontation on the subject may be punished for his impudence by being mocked, snubbed, or made the butt of sneers and ridicule. Some of the terms used are 'idealist', 'pacifist', 'moralist', 'holier-than-thou'. Humphrey concludes that the Bomb is not an uncontrollable automaton, and that we are not uncontrolling people. Our control lies in the force of public argument and public anger.

Civil Defence

In the face of massive casualties from a nuclear attack no rational medical response can be planned. This applies to mass evacuation, as well as to the futile hope that shelters might help prevent delayed damage to survivors from radiation. In fact because the concept of building fallout shelters raises false hopes of protection against the ravages of nuclear war, it is now considered that medical practitioners who advocate the building of nuclear shelters are guilty of gross medical negligence.

Prevention of Nuclear War

The only rational response to the massive medical problems which the threat of nuclear war poses is to ensure that nuclear war never occurs. This solution can only be achieved if the world progresses rapidly to nuclear disarmament.

With this in mind physicians are increasingly campaigning in the political arena, and are joining the growing protest throughout the world. International Physicians for the Prevention of Nuclear War Inc. (IPPNW) was founded by a US and a Soviet physician and the first congress was held in Virginia in March 1981. There were 73 formal delegates from eleven countries including 13 from the USSR. The second conference was held in Cambridge, in April 1982, and the third was in Amsterdam in June this year. In June 1982 an unscripted hour-long round table discussion between three US and three Soviet members of IPPNW, was broadcast twice on Soviet television and seen by an estimated 100 million people. The Physicians for Social Responsibility in America has grown to more than 18 000 members. The Australian Branch of the Medical Association for Prevention of War (MAPW) also is growing rapidly in membership.

The medical profession has a special role to play in preventing nuclear war, because of the intimate knowledge that its members have of death and human suffering.

References

British Medical Association (1983), *Report of the Board of Science and Education into the medical effects of nuclear war,* British Medical Association, London.

Humphrey, N. (1981), 'Four minutes to midnight', *Listener,* 29 October.

Middleton, H. (1982), 'Epidemiology: The Future is Sickness and Death', *AMBIO, 11,* 100.

5 CAN WE SURVIVE A NUCLEAR ATTACK UPON AUSTRALIA?

J. A. Ward

J. A. Ward

The importance of the United States military bases in Australia make it almost inevitable that our country will be attacked in a major nuclear war between the superpowers. Such an attack would certainly be directed at the US bases of North-West Cape, Nurrungar and Pine Gap. It is likely also that HMAS Stirling at Cockburn Sound, used by US nuclear missile carrying submarines and surface ships and the RAAF base at Darwin, used by the US B-52 bombers would be attacked if in use by US forces at the time, or even to prevent their use by US forces. The main uncertainty is whether other population centres in Australia are targeted and it is the strongly held view of the Medical Association for Prevention of War (MAPW) that this issue has been inadequately debated. On balance, I find the argument that our major ports would be bombed to prevent our supplying the US with desperately needed fuel and other resources during the post-war period, no less persuasive than the counter argument that the USSR does not have enough warheads to waste a dozen on Australian cities. Even if the latter is currently the case, it may not be so in five or ten years.

It is therefore the view of MAPW that Australia should expect an attack on the US bases, Cockburn Sound, Darwin and the port areas of Sydney, Wollongong, Newcastle, Melbourne, Geelong, Adelaide and Fremantle if an all out nuclear war begins between the USA and the USSR. The deaths and injuries from such an attack would be an unimaginable catastrophe, and the most important task facing all Australians today is to make sure that it never happens.

One of the points on which doctors throughout the world agree absolutely is that the health services will be incapable of making any sort of effective response to the casualties of a nuclear attack. Governments should understand without doubt that the injured survivors of a nuclear attack will live or die virtually untouched by modern medicine. There will be no health services to diminish the guilt of those who are leading us to the nuclear holocaust, the 'final epidemic'.

Let us consider, for example, the case of a single 1 Mt bomb exploded at ground level in the centre of Sydney. At the same time, we should be aware that a Russian attack on the port areas and oil-refining facilities of Sydney could entail more than one bomb. A single 1 Mt bomb, however, would kill up to 200 000 people instantly and leave up to 300 000 severely injured (Smith and Smith, 1981; United States Congress, 1979; Humphreys, Hartog and Middleton, 1982; Andrews, Powles and Ward, 1983). The injuries would be commonly multiple and include crush injuries to the head, chest and abdomen; fractures and lacerations. There would be between 10 000 and 50 000 serious burns depending on the time of day, the weather conditions and the warning time available. Many of the injured would be also burned and many of both the burned and the injured would have suffered significant radiation exposure. If, at the time of the ground-burst explosion, the wind was blowing from the north-east, east or south-east, between 500 000 and 1 000 000 people could be exposed to lethal doses of fallout radiation. Given the current situation, with virtually no fallout shelter protection for the general public, one must assume that the bulk of this exposed population would suffer severe radiation sickness, much of which would end in death.

Map 1 shows the likely extension of blast overpressures of 12 psi and 5 psi following a 1 Mt bomb explosion in the centre of Sydney. Map 2 superimposes on this the likely area that would be exposed to lethal levels of fallout radiation following a 1 Mt ground-burst in the centre of Sydney, with a 10-15 kph wind blowing from the east.

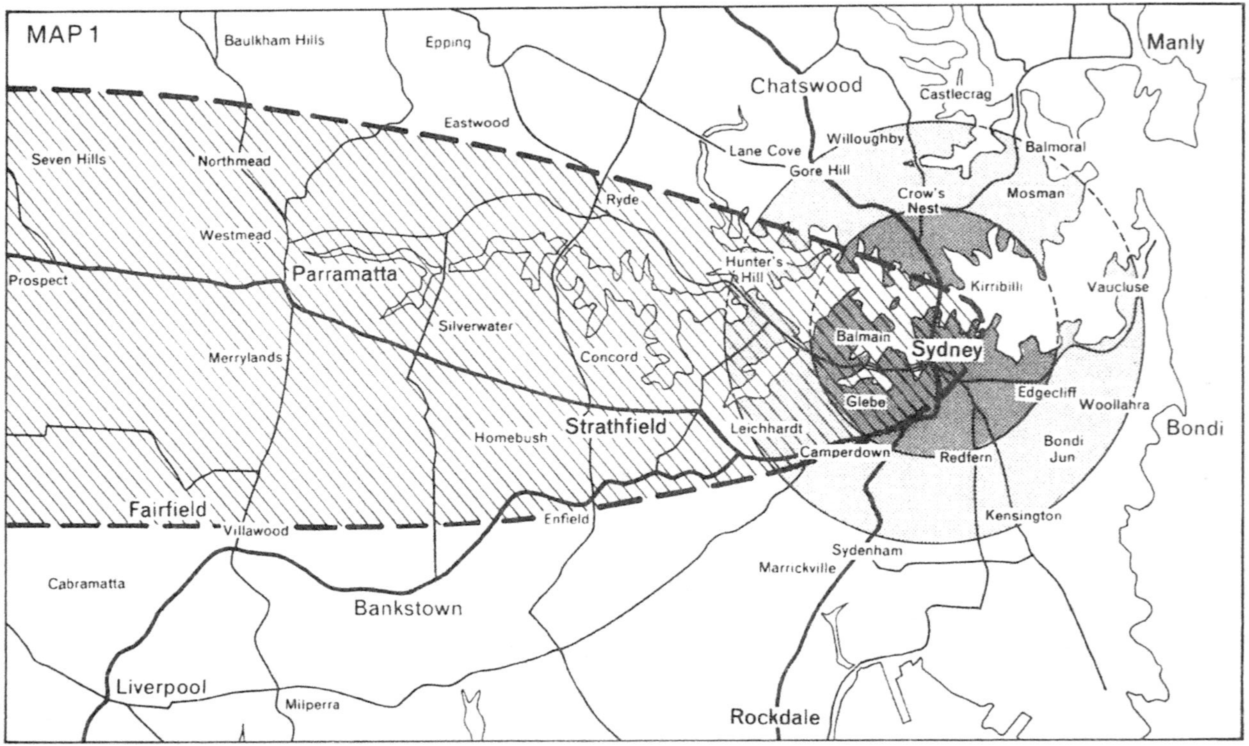

12 psi and 5 psi limits following 1 Mt bomb on centre of Sydney.

Area of exposure to lethal fallout radiation following 1 Mt bomb on centre of Sydney (with 12 psi and 5 psi limits superimposed).

Sydney is no different to most major cities in having its hospital beds and medical facilities disproportionately concentrated in the inner suburban areas. Table 1 shows the number of hospital beds in each of the zones outlined in Map 2 (NSW Health Department, 1982).

For the purpose of this presentation I am assuming that all hospitals, public and private, within the 3.5 km radius are damaged to the extent that they are unusable, but that all public hospital beds outside the 12 psi range survive in a usable condition. I am assuming, however, that private hospitals are less solid in construction and more susceptible to fire and that 50 per cent of private beds in the 5-12 psi area are destroyed. All beds in the lethal fallout area would be intact but unusable for at least two weeks. The total number of hospital beds in Sydney that would not be available to care for the casualties from a nuclear attack are listed in Table 2.

To cope with up to 300 000 blast injuries, 10 000 to 50 000 burns victims and up to 500 000 people suffering radiation sickness, Sydney will have available only 50 per cent of its hospital beds, i.e. 6800 public hospital beds and less than 2000 beds in private hospitals. Unfortunately, these beds will not be empty at the time of the nuclear attack and one wonders what will happen to the existing occupants.

Assuming that doctors and nurses during working hours are geographically distributed in the same pattern as are hospital beds, then a nuclear attack during working hours could kill up to 50 per cent of Sydney's doctors or nurses. More doctors and nurses would presumably survive if the attack was outside working hours or if

TABLE 1

	HOSPITAL BEDS			
	DISTANCE FROM GROUND-ZERO IN KILOMETRES			
	0-3.5 (> 12 psi)	3.5-6.0 5-12 psi	> 6.0 (< 5 psi)	> 3.5 but within lethal fallout area
PUBLIC	3084	2964	6401	2572
PRIVATE	332	993	3009	1518
TOTAL	3416	3957	9410	4090

there were sufficient warning to allow people to return to their homes or to stay at home. One horrendous possibility is that the warning time is 15-30 minutes, i.e. the travelling time of submarine-launched missiles and land-based ICBMs and people are caught in the traffic jams trying to reach their homes.

Let us assume, however, that 50-60 per cent of Sydney's doctors and nurses have survived the immediate effects of the bomb and are not trapped in high-fallout areas. We know from the recent disaster at Three Mile Island (Maxwell, 1982) and the war in Lebanon that up to 50 per cent or more of these doctors and nurses will not report for duty. They will be doing what everybody else is doing — fleeing the city, looking for missing relatives, coping with damaged homes or injured family members. In one major hospital near Harrisburg only six doctors out of seventy reported for duty.

Let us assume, therefore, that 30 per cent of Sydney's 5000 doctors report for duty. Of these, probably half will be of little more use than laymen because of their lack of recent clinical experience. This leaves 800 useful doctors to deal with up to 300 000 injured and burned patients. If each doctor spent just 15 minutes with each patient and worked a 16 hour day, it would be six days before every patient was seen once.

Most injured, however, will never see a doctor or arrive at a hospital. The streets will be blocked by fallen buildings and there will be no transport. Many of the more serious injuries and burns will die in the first week without even the comfort of morphine.

Despite the blocked streets and radiation levels, tens of thousands of injured, burned and irradiated victims will arrive at the functioning hospitals. Faced with this impossible task, doctors

TABLE 2

Hospital beds unavailable for use after a 1 Mt ground-burst in the centre of Sydney with a 15 kph wind from the east:

	HOSPITAL BEDS UNAVAILABLE FOR USE			
			TOTAL	
	Destroyed by blast or fire	In high fallout area	Number	% of beds in Sydney
PUBLIC	3084	2572	5656	45
PRIVATE	828	1518	2346	54
TOTAL	3912	4090	8002	48

and nurses will introduce a triage system, the logic of which will be compelling, the inhumanity of which will be abhorrent. All those who should survive without medical care will be turned away as will those who will die despite medical care. The latter group may be taken to a large public building such as a school to be provided with analgesia if stocks allow. Most likely they will die alone and in pain. The only patients admitted to hospital will be those whose survival is judged to be dependant on medical care.

For blast injuries such a triage process will be within the expertise of most practising doctors. It will be considerably more difficult for burns injuries in view of the uncertainty as to survival in the absence of the massive levels of resources normally devoted to these patients. For radiation sickness, it will be virtually impossible because the early symptoms are similar regardless of whether the person has been exposed to a survivable or non-survivable dose. In the range of 100 to 4000 rads the initial symptoms may be anorexia, nausea, vomiting, malaise, diarrhoea and fever. It is not until the fifth day onward that signs and symptoms such as haemorrhage, epilation and swelling of the gums suggest that the radiation doses may be fatal. Redness of the skin within a few hours may suggest a dose of 500 rads or more but the presence of burns will render this sign less reliable. An added complication is that the initial symptoms of vomiting, diarrhoea and malaise will occur also in persons suffering only from acute anxiety states.

The best guide to the exposure dose will be obtained by monitoring the bone marrow through blood tests. Figure 1 shows typical changes in the haematological picture over the first two months following an exposure to a few hundred rads.

As it is the patients who develop the severe neutropaenia and thrombocytopaenia who will benefit from medical care, it is important to identify them and admit them to hospital. The early rise in neutrophilia in the first 48 hours and the severity of the early lymphopaenia are the best indications of the degree of bone marrow irradiation. The problem, of course, is whether such intensive haematological monitoring will be available.

Despite the difficulties facing the medical staff in deciding which of the tens of thousands of patients will benefit most from admission, we can be sure that the hospital will soon be overflowing with patients in great pain suffering from severe injuries, burns and radiation exposure. The resources, both human and non-human, will be totally inadequate to meet this task. If Sydney is attacked,

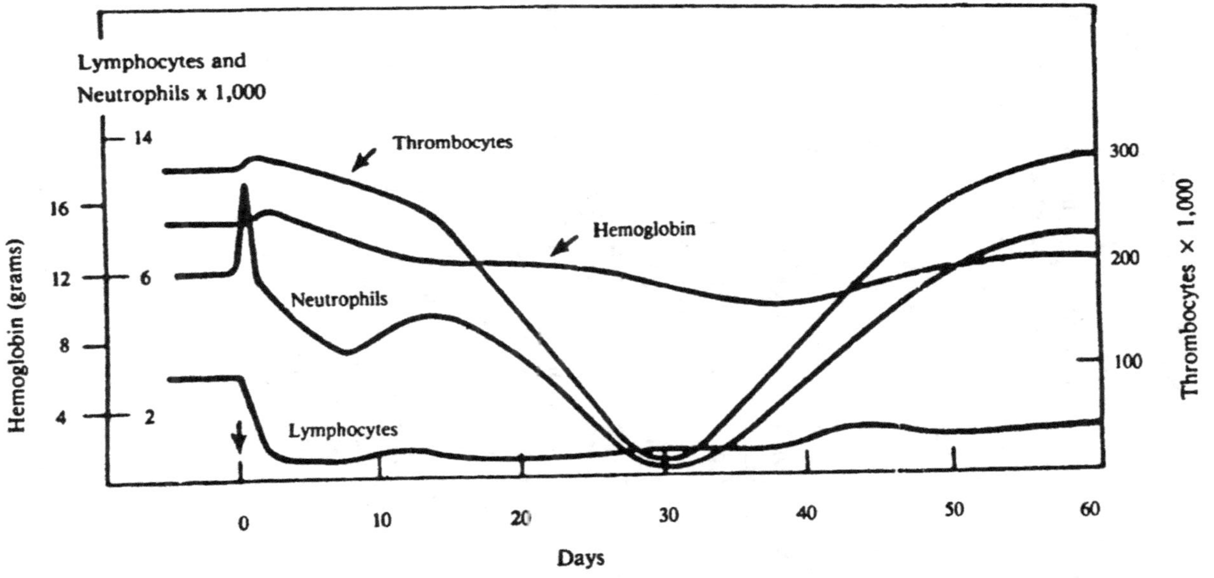

FIGURE 1 (Lindop and Rotblat, 1981)
Typical haemotological changes following exposure to 300-600 rads

one can assume that Newcastle, Wollongong, Melbourne and possibly Canberra will have been attacked as well. Unlike the situation in Hiroshima and Nagasaki, there will be no inpouring of resources from outside. In post-attack Sydney, there will be desperate shortages of manpower, matched quickly by desperate shortages of drugs and surgical supplies. There will be insufficient operating theatres, anaesthetists and surgeons to cope with the crush injuries and compound fractures. For the burns patients there will be no sterile conditions, insufficient topical antiseptics and dressings and most importantly, insufficient blood and plasma. A person with 20-30 per cent second or third degree burns can require 20-30 litres of blood to survive. When I rang the Red Cross Blood Bank to find out how much blood and plasma we had stockpiled in Sydney, I was told that they did not know but that it did not matter as people would flock in to give blood in the first few days after a nuclear attack.

It is obvious that a hospital staff at about 25 per cent of its normal level, faced with thousands of severe injuries and burns will be able to offer only the barest minimum of medical and surgical care. Laboratory services will be completely overtaxed and may not function at all if key personnel fail to report for duty. The standard of care will be less than that provided in a front-line battle situation and may be more akin to that of a century ago, when, in the absence of the trappings of modern medicine, human skill was the main resource.

The hospitals will remain overflowing with bomb victims for many weeks as the consequences of trauma and burns unfold and as those exposed to high radiation doses move into the phases of infection, bleeding, anaemia and gastro-intestinal complications. During these months, the routine morbidity of the population can receive no attention. More importantly, the devastating psychological effects of the bomb will progress without intervention.

Civil Defence Against Nuclear Attack

From their analysis of nuclear weapons and the likely effects of nuclear war, many doctors in Europe and North America have concluded that civil defence measures against nuclear attack are of little use. Moreover, in view of the possibility that such measures

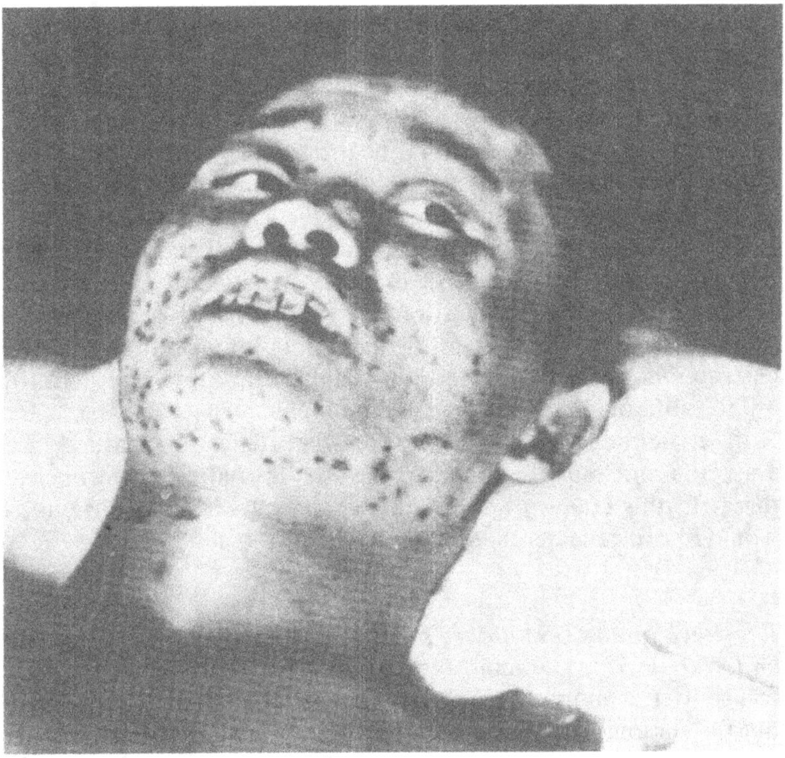

Hiroshima victim with bleeding under the skin of his face and in his gums, and loss of scalp hair due to radiation injury. He died 3 September 1945

condition the public to accept the arms race and the consequences of that race, some doctors have declared that any medical practitioner who participates in civil defence planning is committing a profoundly unethical act. It is their view that the involvement of doctors in such exercises lends credence to the illusion that civil defence measures will protect the public against nuclear attack, thereby diminishing the unacceptability of nuclear war.

It is my view that this analysis is appropriate for countries which have nuclear weapons or which base them on their soil and could thereby expect to receive saturation bombing in a major nuclear war. I do not believe it is appropriate for neutral or non-nuclear-weapons countries in Europe nor for Australia. Countries such as Sweden, Norway and Switzerland will, presumably, not be targets for nuclear attack and will therefore suffer mainly from fallout radiation in the post-attack period. In the long term, of course, they may suffer a whole host of catastrophic ecological and social effects from which there may be little useful defence. These countries have embarked on major programmes of fallout shelter construction, to the extent that the bulk of their population would be protected from fallout radiation in the event of a nuclear war. For the USA, USSR, UK and Germany, however, bombing will be intensive and survival will be as much a matter of chance as of preparedness. It is also questionable what survival will mean in a social sense, with 50 per cent of the population dead, almost total destruction of industry and oil-refining facilities and tens of millions of homeless, starving, contaminated refugees.

In examining the usefulness of civil defence measures for Australia, one must first analyse the possible nature of the nuclear threat to this country and its likely consequences. To this end, I would like to consider three possible scenarios.

1. Attack on the US military bases of North West Cape, Pine Gap and Nurrungar. The main target on North West Cape, the VLF transmitter is about 15 km from the town of Exmouth (pop. 1500) but the station headquarters and the high frequency transmitter are only about 5 km from the town. If the transmitter only was hit, Exmouth would be spared severe blast and heat damage but may have to cope with fallout radiation. If, on the other hand, the station headquarters were attacked, Exmouth would be more or less destroyed.

The other major towns within the range of lethal fallout are Learmonth (pop. 300), 50 km to the south, and Onslow (pop. 400), 100 km to the west.

Pine Gap is 20 km south-west of Alice Springs (pop. 15 000) and clearly the worst case secenario would be a ground burst hit on Pine Gap, with the wind blowing from the south-west.

Nurrungar is 500 km north-west of Adelaide, only 150 km from Port Augusta and less than 250 km from Port Pirie and Whyalla. A direct attack on Nurrungar with the wind blowing from the north-west would contaminate these three towns but probably with sub-lethal doses of radiation.

We can therefore estimate the deaths that could be produced by the worst case attack on these three US military installations.

TABLE 3

Deaths from worst case attack on North-West Cape, Pine Gap and Nurrungar

	DEATHS	
	From Blast and Heat	From Fallout
North-West Cape	1 500	5 000
Pine Gap	120	15 000
Nurrungar	1 000	2 000
TOTAL	2 620	22 000

2. *Attack on RAAF Base, Darwin and HMAS Stirling, Cockburn Sound.* Cockburn Sound is 20 km south of Perth which would therefore be exposed to lethal doses of fallout radiation if the wind were blowing from the south. The blast effects would certainly extend into the urban area of South Fremantle.

An attack on RAAF Base, Darwin, could devastate the city from blast and heat effects of the bomb.

The possible consequences of worst case scenarios are shown in Table 4.

TABLE 4

Deaths from worst case attacks on Cockburn Sound and RAAF Base, Darwin

	DEATH	
	From Blast and Heat	From Fallout
Darwin	40 000	10 000
Cockburn Sound	20 000	250 000
TOTAL	60 000	260 000

3. *Direct attacks on the ports of Sydney, Newcastle, Wollongong, Melbourne, Geelong, Fremantle and Adelaide.* The deaths from worst case scenarios assuming one single 1 Mt bomb for each port area are given in Table 5.

Finally, if we consider the possibility that the US bases, the Australian bases used by US nuclear weapons carrying submarines and bombers and the major Australian port areas are all attacked, the possible consequences could be of the order of 1 000 000 deaths from blast and 3 000 000 deaths from fallout.

TABLE 5

Deaths from worst case 1 Mt attacks on each of Sydney, Melbourne, Geelong, Newcastle, Wollongong, Fremantle and Adelaide

	DEATHS	
	From Blast and Heat	From Fallout
Sydney - Port Jackson	200 000	500 000
Sydney - Port Botany	80 000	500 000
Newcastle	100 000	200 000
Wollongong	80 000	100 000
Melbourne	200 000	500 000
Geelong	100 000	200 000
Port Adelaide	100 000	300 000
Fremantle	100 000	300 000
TOTAL	960 000	2 600 000

To significantly reduce the number of deaths from the direct effects of a nuclear attack (blast and heat) would require the construction of blast shelters to withstand 40-80 psi of blast overpressure. At an average cost of $20 000 per family they would be beyond the range of most families and their ability to protect their occupants must be doubtful. Moreover, many families living within 5 km of port areas live in rented accommodation and would be dependent on public shelters. I understand that the Fallout Shelter Survey Team of the Natural Disasters Organization has surveyed the capital cities and identified places suitable for public fallout protection (Posener, 1982).

Apart from the planned shelter beneath the new Parliament House in Canberra, I know of no public blast shelters in Australia and I do not believe that such a programme would be cost-effective. It is therefore inevitable that a nuclear attack upon Australia involving the major port areas will result in at least 1 000 000 deaths regardless of civil defence planning and the politicians and public alike must be made aware of this fact. It represents such a catastrophe that it must never be allowed to happen.

Having acknowledged and understood the irreducibility of these 1 000 000 deaths, it is then important to realise that an additional 3 000 000 deaths could result from fallout radiation and could theoretically be prevented by a complete programme of fallout shelter protection. Fallout shelters need to have a protection factor (PF) of 40 or more, i.e. screen out at least 97.5 per cent of radiation, requiring walls of about 40 cm of concrete or about 60 cm of dirt (Ball, 1982). As it will be about 2 weeks before people can emerge from their shelter for long periods at a time, the shelters will need to be habitable for such a period, requiring water, food, filtered air and toilet facilities. Family shelters would be the least unpleasant and at approximately $10 000 per shelter would be within the capacity of many families. To cover families in rented accommodation and persons at work, a range of public fallout shelters would also be required. Government financial assistance will obviously be required for low income families. One of the more unpleasant aspects of fallout shelters is that families would presumably have to repel, by force ultimately, unwanted would-be invaders. The prospect of killing a neighbour to ensure the survival of one's family is enough in itself to compel us to campaign urgently for nuclear disarmament.

Finally, mass evacuation offers some hope and is the current main focus of interest for the NSW State Emergency Services. As far as I know, they have no formal plans for a co-ordinated evacuation, but are simply relying on up to 1 000 000 Sydney residents leaving the city during the weeks before the attack, in which they presume hostilities between the superpowers are building up. Planned mass evacuation would have a number of problems and uncertainties:

- would there be sufficient warning to give people time to leave the city?
- people caught on the road would receive far more radiation than those who stay at home.
- would people obey the rules in relation to leaving by road or will there be one giant traffic jam?
- would people respond to a call for evacuation if there had been a couple of false alarms?
- would drivers of public transport return to the city to pick up further people or would they simply disappear into the fleeing crowds?

I do not believe that medical disaster planning will make any significant difference to the ineffectiveness of the health service response. Despite such planning and annual practice exercises we have struggled to cope with the disasters that have occurred in the last decade or two. It is simply not in man's nature to prepare for an event that may never happen. More importantly, no amount of planning and training can prepare us to cope with 300 000 severe casualties.

The peace movement in general is strongly opposed to civil defence planning — in part because it suggests an acceptance of the inevitability of nuclear war; in part because it projects a belief that nuclear wars are survivable, but also because of the socio-political issues raised by civil defence. The peace movement believes that the politically and financially powerful will ensure their own protection while that of the public is accorded a lower priority. Such a state of affairs seems particularly unjust to the public who see the politically and financially powerful as the very people who are promoting the arms race and obstructing the movement for nuclear disarmament. To overcome the reservations of the peace movement, the State and Federal Governments would have to be seen to be promoting disarmament and peace by supporting the freeze movement, particularly in the United Nations, by establishing a Ministry for Peace and a Peace Research Institute and by working towards a nuclear-free Pacific region. In addition, the Governments would have to provide generous assistance for low income earners to construct shelters, and finance the construction of adequate public fallout shelters for people in rented accommodation and near workplaces.

Clearly civil defence is no solution to a nuclear attack. Even if the entire population were provided with fallout shelter protection, a nuclear attack on the major port areas will leave 1 000 000 dead and an equivalent number of severe injuries and burns. The only effective defence against nuclear attack is to either ensure that a nuclear war does not occur by means of multilateral disarmament or to try to ensure that Australia is not a target if nuclear war does occur. The latter would entail dismantling the US bases in Australia, refusing landing right to nuclear weapons carrying ships and planes and the adoption by Australia of a non-aligned defence strategy. I am not a defence strategist but if there was a medical problem for which only one treatment was effective, I would certainly consider it with great interest and I would want some very good reasons for not using it. While I do not have the expertise to pronounce judgement on the US bases and the US-Australian alliance, I feel strongly that, with the stakes so high, the issue has been inadequately discussed and debated to a degree that is almost criminal.

The argument that major port areas in Australia are targeted and would be attacked, along with the US bases, Darwin and Cockburn Sound, to prevent Australia assisting the US in the post-attack period, seems no less convincing to me than the counter arguments. In the event of such an attack, up to 1 000 000 people may die of blast and heat with an equivalent number of severe injuries and burns, while an additional 3 000 000 may die from fallout radiation if the attacks were ground-bursts and the population was unprotected. Facing this incomprehensible number of casualties would be a medical service reduced to 50 per cent of its hospital beds and 20-30 per cent of its doctors and nurses. With hopelessly inadequate levels of blood, plasma and drugs, the level of medical care provided would approximate more to that of a medieval battlefield than to modern trauma surgery. Civil defence can do little to reduce the 1 000 000 deaths from blast and heat but a comprehensive fallout shelter construction programme could save the additional 3 000 000 at risk from delayed radiation. While it is important for this fact to be widely known, the consequences of a nuclear attack, despite fallout shelter protection, would be so catastrophic as to be totally unacceptable. I would not like to see the promotion of civil defence sapping our political energy which must be directed predominantly towards ensuring that nuclear war does not occur. The immediate requirement is a freeze on the development, testing and deployment of nuclear weapons followed by multilateral, verifiable reductions in all weapons systems, while at the same time seeking methods of solving international disputes without war.

References

Andrews, J., Powles, J. and Ward, J. (1983), *Medicine and Nuclear War,* Victorian Association for Peace Studies, Peace Dossier 5.

Ball D. (1980), *A Suitable Piece of Real Estate:* American Installations in Australia, Hale & Iremonger, Sydney, p.51.

Ball, D (1982), Limiting Damage from Nuclear Attack, presented at Conference on Civil Defence and Australia's Security, ANU, April.

Humphreys, J., Hartog, M. and Middleton, H. (1982), *The Medical Consequences of Nuclear Weapons,* Medical Campaign Against Nuclear Weapons, London.

Lindop, P. J. and Rotblat, J. (1981), 'Consequences of Radioactive Fallout' in R. Adams, S. Cullen (eds.) *The Final Epidemic,* Educational Foundation for Nuclear Science, Chicago, p.145.

Maxwell, C. (1982), 'Hospital Organizational Response to the Nuclear Accident at Three Mile Island: Implications for Future-oriented Disaster Planning', *American Journal of Public Health, 72,* 275-9.

NSW Health Dept., Division of Planning (1982), *Hospital Planning Data,* May.

Posener, D. W. (1982), 'Civil Defence and Fallout', presented at Conference on

Civil Defence and Australia's Security, Strategic and Defence Studies Centre, Research School of Pacific Studies, ANU Canberra 19-22.4.82.
Smith, J. and Smith, T. (1981), 'Nuclear War: the medical facts', British Medical Journal, 238, 771-774.
United States Congress, Office of Technology Assessment (1979), *The Effects of Nuclear War,* Washington, DC, US Government Printing Office, (OTA-N5089).

6 DARK CIRCLE
Judy Irving and Heather Ogilvie

Judy Irving

To write about a film in such a way that the integrity and power of that film is not lost, is a daunting task. But when the message of the film is as vital as that of *Dark Circle,* it is imperative that its story be told.

Five years in the making, *Dark Circle* is a haunting portrait of the nuclear age told through the lives of those directly affected by it, in an interlocking series of 'atomic biographies.'

Dark Circle was shot on location at the Rocky Flats Nuclear Weapons Facility near Denver, the Diablo Canyon Nuclear Power Plant in central California, Hiroshima and Nagasaki, Japan and in the Aleutian Islands of Alaska. The film graphically shows what building, testing, selling and using hydrogen bombs means in individual, human terms and how it is touching the lives of ordinary people — even in the absence of a nuclear war.

Dark Circle makes extensive use of archival footage, much of which has been declassified or uncovered for the first time. In some cases this involved two years of letters, phone calls, on-site visits and eventually congressional pressure to obtain release of the relevant film.

The film was completed in October, 1982 and had its world premiere at the New York Film Festival, where it received the only standing ovation of the event.

Much of the material in this chapter has been taken directly from the script of *Dark Circle* and each section examines a different theme running through the film. Through the courage and actions of the people presented in the film, *Dark Circle* forcefully demonstrates that there is still hope, the choice is ours. It may not be too late.

The Relationship of Nuclear Power to Nuclear Weapons

Narrator . . . this is where plutonium is created, in nuclear reactors — in every nuclear reactor. It's the only place plutonium is made. Nuclear reactors are the birthplace of plutonium . . . a reactor of this size will produce a thousand megawatts of electricity and every year five hundred pounds of plutonium. This material is commonly called 'radioactive waste' but its also the raw material for atomic bombs. A plant the size of Diablo Canyon produces, within its radioactive core, enough plutonium every year to build fifty nuclear weapons.

Dark Circle simultaneously explores the unbreakable links between nuclear weapons and nuclear power, links which are studiously played down by the nuclear industry: from the creation of plutonium as an inevitable by-product of nuclear power for electricity to plutonium's use as a fuel for bombs.

Pam Solo is a Catholic nun and a veteran organizer, working with the American Friends Service Committee.

Pam Solo Nuclear reactors became a nursery for those who would build bombs, and so the Soviet Union exported nuclear power reactors to China and China soon joined the nuclear weapons club. Ironically now China's nuclear weapons are all pointed at the Soviet Union. The United States and Canada sent nuclear technology and advisors to India. India joined the nuclear weapons club. So the reactors themselves are a link in, a

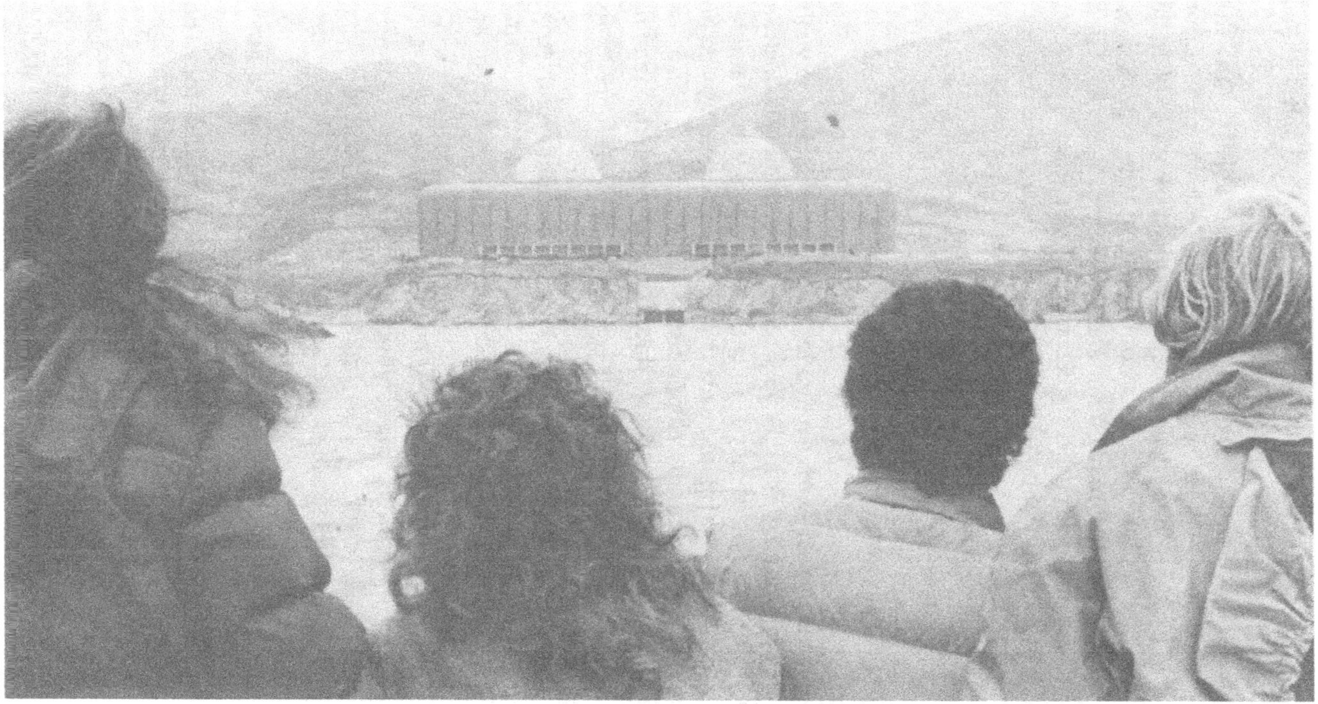

Four kilometres from an active earthquake fault, California's Diablo Canyon Nuclear Power Plant
© *Karen Spangenberg*

road into the nuclear weapons club and without them you can't join.

Narrator . . . and while we're preoccupied with the Russian threat, our own export of nuclear technology has begun to backfire. In 1974 India performed an astonishing feat, the explosion of an atomic bomb, made entirely of materials from their peaceful nuclear power program. The secret of the bomb was no longer a secret and the most direct route to nuclear weapons was nuclear power.

The Hidden Face of the Nuclear Industry

As the production of *Dark Circle* illustrated, penetrating the walls of the nuclear industry is never easy. Nuclear weapons production has been shrouded in secrecy for nearly forty years with 'national security' serving as a blanket excuse to keep it that way. The filmmakers viewed literally hundreds of thousands of metres of film in thirty archives from Washington DC to Nagasaki, Japan in order to assemble a comprehensive look at the hidden world of nuclear weapons.

The film delves into the actual working of the miltary-industrial complex at the private annual arms convention of the US Army Association, where new weapons, both nuclear and non-nuclear are sold to the military.

Narrator The same companies that build the reactors that produce plutonium also build the hydrogen bomb . . . the military-industrial complex is a well known fact of life, but the seven corporations and one university that build the hydrogen bomb prefer not to advertise their involvement.

Bendix produces packaging material for The Bomb. Dupont supplies radioactive hydrogen gas. A subsidiary of American Telephone & Telegraph performs engineering studies. General Electric manufactures a neutron generator to ignite the plutonium trigger. Union Carbide provides enriched uranium for the hydrogen

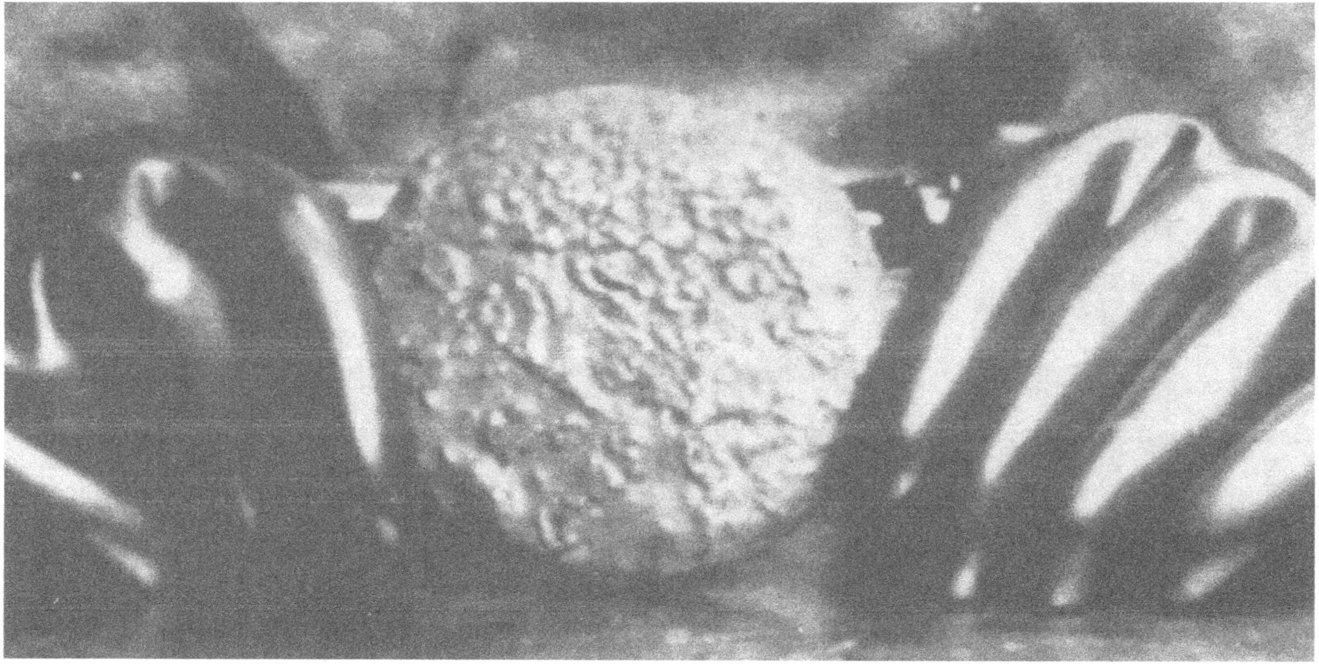

This is a plutonium button, weighing four pounds. It contains enough particles of plutonium to give cancer to the entire human race, if it were evenly divided and inhaled by each person. *[from 'Dark Circle']*

bomb. Monsanto manufactures non-nuclear explosive detonators. Mason & Hanger Silas-Mason assemble The Bomb in Texas. The University of California designs new weapons like the neutron bomb. But the heart of the nuclear weapons complex, where Rockwell International oversees the production of the plutonium trigger, is Rocky Flats.

Three plutonium triggers a day, fifteen a week, seven hundred and twenty bombs a year, many of them hundreds or even thousands times more powerful than the bombs dropped on Japan. Twenty-six thousand hydrogen bombs stockpiled, waiting for the next war . . . the profits from the manufacture of these weapons, due to cost-plus contracts and cost overruns, average three times as high as the profits on consumer products.

Recorded Voice (Rockwell Booth Arms Convention) On the battlefield, swift and deadly, a formidable foe. The Warsaw Pact continues to strengthen its armed forces as well as other offensive weapons, while proclaiming its intentions for peace.

Narrator When fear of the Russians turns up as a sales technique at the Rockwell Booth, its hard to distinguish the real threat from the advertising.

Dark Circle gives a face to the hitherto secret world of nuclear weapons where the mythic bomb of the past can be finally seen as the manufactured product of today. *Dark Circle* begins to lift the cloud, not only on the manufacture of nuclear weapons but also on our ability to mobilise by giving us these faces.

Contemporary Effects of Nuclear Weapons

Marlene Batley lives five and a half kilometres from the heart of the nuclear weapons industry. She was not told when she moved there with her husband and two children, that their backyard was contaminated with plutonium — the residue routinely released as part of the manufacture of hydrogen bombs.

Marlene . . . he said there was a problem with the soil

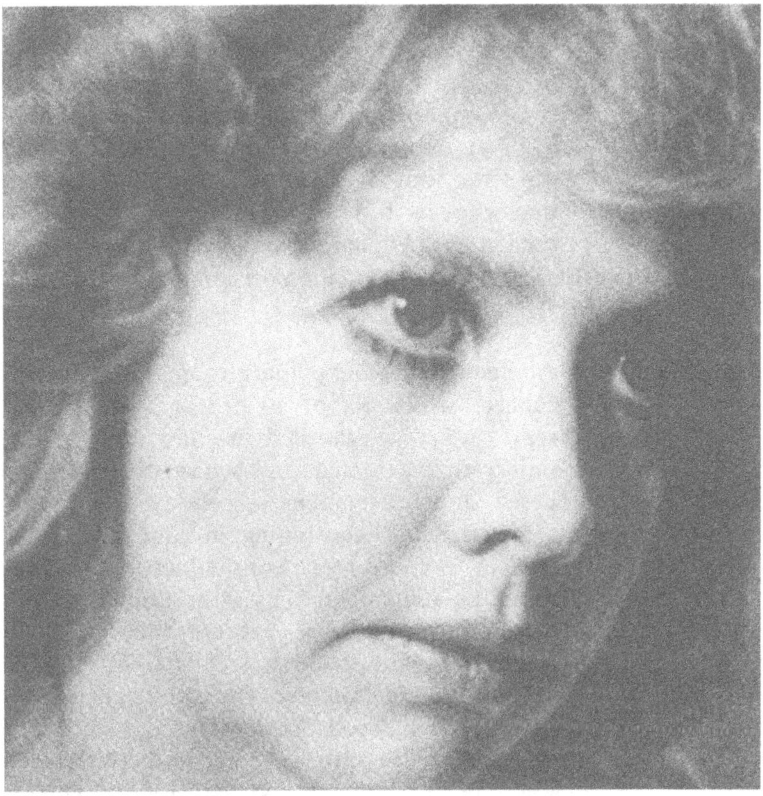

Marlene Batley

sample, however he never said anything about contamination . . . we were never told that.

Marlene 'You can't see it, you can't feel it but it can give you cancer or leukaemia in years to come and you'd never be able to prove it.'

Marlene's growing awareness of the cancer and genetic hazards posed by plutonium throw her into a moral crisis: should she stay and fight the plant or, for her daughters' sakes, try to sell her contaminated house? If she sells, does she inform the prospective buyers of the dangers posed by Rocky Flats?

Marlene I feel no amount of security anymore living in this house, or walking around in this neighbourhood or breathing this air or working in this dirt. I've lost any security that I had when I first moved here . . . now I'm trying to sell I didn't want anybody to say anything about Rocky Flats or nuclear power plants . . . I was afraid something they would say would cost me the sale of my house . . . I found myself with the shoe on the other foot, I guess you'd say . . . I just hope that whoever buys it, maybe they're going to be the kind of people who aren't concerned . . .

For Marlene Batley, the result of the decision she was to make would bring her full circle.

Pam Solo A place like Rocky Flats represents not only cancer, which all of us fear . . . but it also represents, like, the end of some kind of continuity that . . . that we all need to be remembered or to remember somebody and to find some sense of continuing in time and space; nuclear weapons represent the fact that you may never be remembered by your family or your children or whatever because they won't be there . . . its the end.

Narrator . . . and people still feel the effects of the first bomb ever dropped. Yuriko Hatanaka is a child of the atomic bomb. . . . Yuriko is thirty seven years old, with the mental capacity of a child of two. Mrs Hatanaka was three months pregnant with Yuriko when the bomb dropped a mile away. Mr Hatanaka, a barber, was away from Hiroshima that day. Six months later Yuriko was born, and thirty two years later her mother died of bone cancer, another official casualty of The Bomb. Every year since 1945 several hundred people die from the long-term effects of the bomb that destroyed Hiroshima in seconds. For Yuriko the effects will last her a lifetime.

Richard 'Mac' McHugh was eighteen years old when he flew through a glowing atomic cloud in the south Pacific during an atomic test.

'Mac' McHugh The bombs being used against Americans. That's what it was used as in the Pacific; it was used

Richard 'Mac' McHugh

against the Japanese twice but they've exploded a hundred of them out there in Nevada, the Pacific; civilians, even civilians are involved and, I don't know, maybe we're guinea pigs, maybe that's the way they're gonna find out. I don't know.

On film McHugh recalls, 'it was no big deal at the time, but its a big deal now. . . .' He later developed leukaemia and like thousands of other atomic veterans (some 300 000 people took part in atomic tests) McHugh has been denied financial compensation by the Veterans Administration.

Narrator . . . in over seven hundred nuclear tests in Nevada, New Mexico, Colorado, Alaska, Mississippi and the Pacific Islands of Eniwetok and Bikini, hundreds of Americans have seen the bomb and they were told it was safe. Over 2000 atomic veterans have filed for disability compensation . . . the US Government has granted twenty four.

For many of the residents of Rocky Flats, the Nuclear Weapons Facility provides a source of secure income, at a time of high unemployment.

Nancy Wood (resident) It throws you into a great deal of turmoil, because on the one hand we need a job that paid a decent wage, and on the other hand we were living off of literally manufacturing bombs. It's a high price to pay.

In return for that security the workers of the Plant are required to handle thousands of pounds of plutonium as they assemble the triggers.

Don Gabel was one such worker, employed by Rockwell to 'work plutonium' from the age of twenty for nine and a half years.

Don Gabel Working in dry box, gloves, where the gloves would get holes in it you know, get rotten, just being in there getting hot. They had this pipe right by my head where I had to purge them, it was very radiation hot, it was a real high level of radiation. My head would be right by the pop . . . by the pipes and I, I would tell . . . ask my boss about that, he'd say, 'It wouldn't matter your head it's your body you have to worry about.' So . . . I got some inside me, and all over, outside me and I've inhaled some of it . . .

Don Gabel died at the age of thirty of a malignant brain tumour. At the time of going to press, Kae Gabel is suing Rockwell International for her husband's untimely death.

Raye Fleming: The Story of an Activist

Raye Fleming, a young mother with two children, has been fighting for ten years in California, trying to prevent the impending licencing and operation of the Diablo Canyon Nuclear Power Plant only a few kilometres from her home. *Dark Circle* followed Raye's

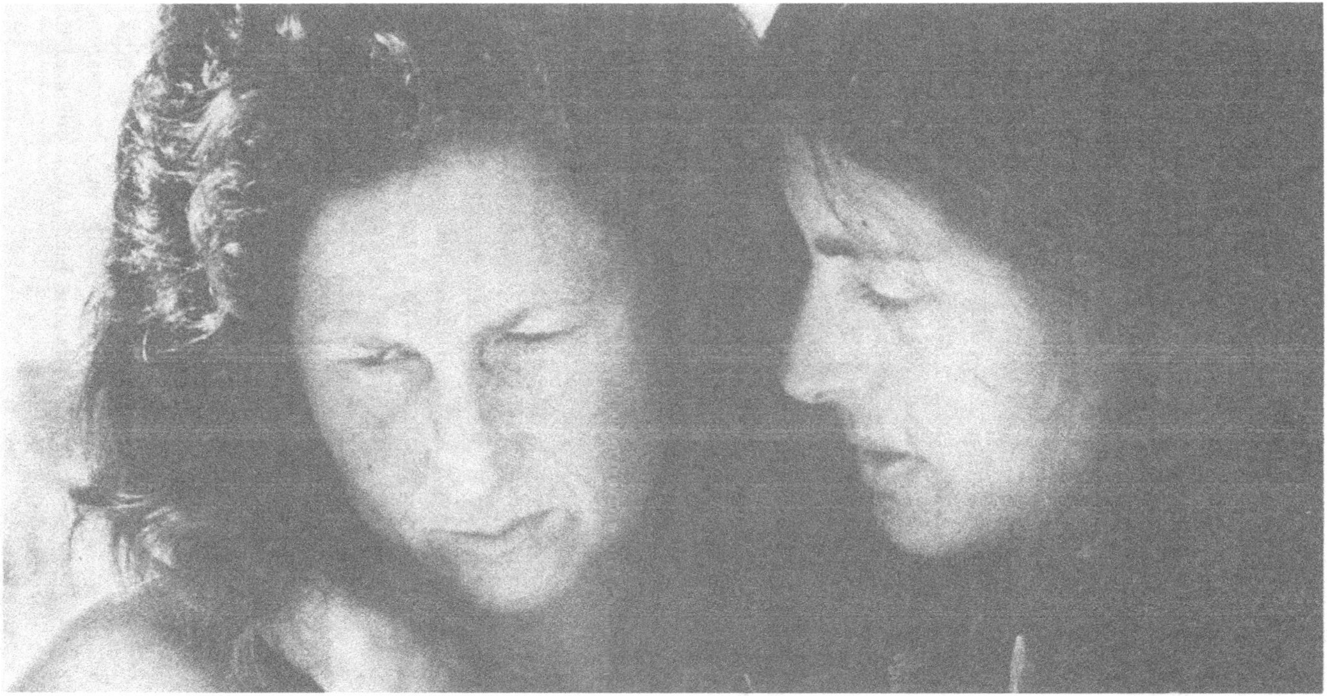

After ten years of legal battles and non-violent blockades, Raye Fleming (left) and a fellow activist share a quiet moment

© *Karen Spangenberg*

story over a four year period, from her initial attempts at legal intervention to her role in helping organize a massive non-violent blockade of the plant itself. Here the film vividly demonstrates the impact one person can have on the nuclear industry. The following is a transcript of an interview with Raye, parts of which were eventually to be included in *Dark Circle*.

Raye The group that I was in, started in 1971, was the Mothers for Peace. The group formed in opposition to the Vietnam War. And we always worked with non-violence, we always worked with consensus — all the things you hear about in a non-violent direct action campaign — we never talked about it. We never sat around and said: 'In this action we will be non-violent. And this is how we'll do it. Or we need to use consensus cause it has real respect for all people.' We just did it.

Judy Irving So how did you move from opposition to the Vietnam War to Diablo?

Raye Throughout the opposition to the Vietnam War I also opposed the use of nuclear weapons and above ground testing. And kind of slowly — through 1972 — it was a time of awakening for all of us in Mothers for Peace — we had a lot more time to think. We started to realise the connection between nuclear power and nuclear weapons — about the same time we started to look into Breeder Reactors. And then it just clicked one day, that if there were all these problems with the Breeder then possibly, there were problems with Diablo Canyon, which is a fission reactor. So throughout '72 and '73 — as a group — we started researching the nuclear issue. And in the end of '73 we took a stand in opposition to the licencing of Diablo Canyon. And a stand against the use of nuclear power for the generation of electricity . . . basically Mothers for Peace was the first local, grassroots organisation to become legal interveners in the proceedings.

At the Federal proceedings Mothers for Peace discovered certain restrictions. Even though both sides were represented, basic safety questions would never be discussed.

Raye What we couldn't discuss was the fact that there's no way to store the waste from Diablo. We're not allowed to discuss that because it's a generic issue that applies to all reactors. We cannot discuss the fact that the emergency cooling system has failed all of its scale-model tests. So when you're playing their game, they're in total control, and if you look at the history of nuclear power and the Federal Government, you see that every reactor that has been built has been granted a licence to operate. No matter what the objections are, no matter who the experts are that testify that the marine life will die; that there'll be cancer and leukaemia, birth defects; and so you begin to feel that you don't have any power, that you're just going through the motions.

Raye and the Mothers for Peace changed their tactics. During the four years of filming the movement against Diablo increased rapidly, with more and more of the usually conservative element of San Luis Obispo joining the campaign.

A coalition of sixty groups from all parts of California, organised large-scale, highly visible action culminating in a human blockade of the main gates of the plant. The blockade continued, even after the licence had been granted, and became the longest, and in terms of arrests, the largest civil disobedience action in the history of United States nuclear power.

For Raye Fleming, it appeared that her ten year battle had been lost. But at the last moment Pacific Gas and Electric discovered that the cooling systems for the plant's double reactor had been installed backwards. The same cooling system designed to prevent a meltdown. This grave mistake was only the first of hundreds later uncovered in what the Nuclear Regulatory Commission had described as the most highly analysed building in the world.

The Commission had no choice but to suspend Diablo Canyon's licence indefinitely.

For the first time in ten years Raye had some breathing room . . . she and others like her had bought enough time for the mistake to be discovered. After thousands of pages of testimony from the experts, Mothers for Peace had been proven right.

Judy Irving What will you do . . . when Diablo is withdrawn, when the licence is not granted?

Raye Oh I don't know. I wish that Diablo was my only interest. I wish that nuclear power were my only interest . . . I don't think somebody who has ever been involved in the campaign, who's really aware as to what's happening in the world around them, could withdraw after one campaign was finally closed. There's other nuclear facilities operating in California. So I think, probably my work will go on. I'd like to think that women have a lot of power and a lot of strength and that they can do what they want to do.

Dark Circle
An Independent Documentary Group Production

Credits

Produced and Directed by Judy Irving, Chris Beaver and Ruth Landy
Filmed and Edited by Chris Beaver and Judy Irving
Narrated by Judy Irving
Supervising Sound Editor: Karen Spangenberg
First Assistant Editor: Michael Levin
Original Music Score by Gary S. Remal and Bernard L. Krause

Awards

Grand Prize — US Film and Video Festival, 1983
Gold Medal — Houston International Film Festival, 1983
Blue Ribbon — American Film Festival, 1983

Length — 82 minutes, 16 mm colour.
Distributed to all parts of Australia by:
THE SYDNEY FILMMAKERS CO-OPERATIVE,
PO BOX 217, KINGS CROSS, NSW 2011.
(02) 33 0721

7 THE ECONOMIC AND SOCIAL IMPACT OF NUCLEAR WAR FOR AUSTRALIA AND ITS REGION

H. C. Coombs

H. C. Coombs

In 1978 the then Director of Scientific and Technical Intelligence in the Joint Intelligence Organization, R. H. Mathams stated:

> The most significant trend for Australia is the large number of war-heads available to U.S.S.R. which now has sufficient war-heads to adequately target the U.S. and retain substantial reserves for use against secondary targets. We cannot determine the priorities, the U.S.S.R. attaches to targets in Australia but joint U.S.-Australian facilities would probably rank high depending on Soviet perceptions of their strategic importance.

In descending order of probability Mathams concludes Australia might receive strategic nuclear attacks against: US facilities in Australia; Australian Defence Establishments; Industrial complexes and urban centres.

This view is supported by Desmond Ball in his paper for a Conference on Civil Defence and Australian security where he implied that the probability of attacks on US facilities was high.

'There is', he said 'now a widespread acceptance within the Defence community that in any general war between the superpowers there are a small number of targets in Australia which are likely to be attacked. . . . Most analysts would attribute this possibility to Australia's hosting of U.S. Military communications, early warning and intelligence facilities' (Ball, 1983).

He and other strategic analysts believe that the probability of other attacks is not high. The main reason for this judgement seems to be that the USSR would be unwilling to use warheads which they may need to hold in reserve for more important targets or as a means to compel a negotiated settlement (Ball, 1983). The validity of this argument depends presumably upon the size of the reserves of warheads available for secondary targets held by the USSR — which they believe do not greatly exceed the numbers necessary to cover vital US European and Asian targets.

In considering the economic and social impact of the nuclear war between the superpowers on Australia we must take account of:

1. the indirect effects of what has happened in the Northern Hemisphere;
2. the effects of whatever attacks are made on Australian targets;
3. the cost and effects of dealing with these effects;
4. the social and political reactions of the Australian people to these events.
5. the practicability and cost of recovery.

Indirect effects

The Australian economy is closely integrated with the rest of the world — an integration that has been fostered and intensified in recent years by the emphasis in Australian government policies on 'letting the world market forces direct the economy' and by the increased importance, financial, technological and managerial, of trans-national corporations. Indeed there is now no autonomous 'national' economy in any comprehensive sense but an economy composed predominantly of aspects of the world economy located within Australia together with marginal independent components isolated by distance or not yet incorporated into the world economy.

These Australian based parts of the world economy are especially closely integrated with other parts located in USA, Europe and Japan, all of which nation states are likely to be involved in any major superpower nuclear conflict. Thus the major part of Australian imports (almost 70 per cent) derive from the EEC countries, USA and Japan (Australian Bureau of Statistics, 1982). The dependence implied by this ratio, is emphasised by the composition of those imports by far the largest category of which is machinery while petroleum products, transport equipment and chemicals make up, together with machinery, almost 60 per cent of the total. These items constitute a critical part of the capital equipment, the essential materials and the fuels of the Australian industrial system. Without them that system would grind to a halt until alternative sources had been developed. To this aspect of dependence must be added those which arise from reliance upon foreign capital, technology and managerial capacity which, in financial terms is reflected in the 'invisible items' in the balance of international payments (Australian Bureau of Statistics, 1982).

To what degree could Australia count upon the continued supply of these imports of equipment materials and industrial services if our major sources of supply are involved in nuclear war? Arthur M. Katz in a review of the economic and social impacts of nuclear attack on the US written after participating in a study for a US Congress Committee of the Economic and Social Effects of nuclear war on US, concludes that:

> . . . complex inter-dependent relationships make modern industrial society especially vulnerable to nuclear attack. (Katz, 1982, p.138).

In elaboration he judged that after such attacks

> The US economic system will be partitioned and fragmented: its integration as an effective system seriously in question:
> . . . attacks not only damage the industrial base, they thin the ranks of technical and managerial personnel;
> . . . they also destroy and disrupt the supporting financial structure; value and utility of money will be in question, as will the value and ownership of property, and the records of financial transactions; lending and borrowing institutions will collapse;
> . . . casualties will be incomprensible involving loss of skilled

> workers, managers, and economic leadership; producing a staggering social burden of injured and immeasurable costs in loss of social stability.

Where, in the order of priorities of a torn and disrupted US society and its surviving economic and political leaders, would rank the needs of the Australian industrial system?

Commenting specifically on the international implications of attacks on US, Katz (1982) comments:

> Key commodities (especially from the world point of view, food, fertilisers and pesticides exports from U.S.) will be lost to world markets for extended periods.
> Basic international trade and payments mechanisms will have to be restructured.
> Most currencies (including U.S. dollar) will be of questionable value; those countries holding their international reserves in currencies of other countries will face loss of access to them.

While Australia would be less affected than many countries by the reduced availability of US food, fertilisers and pesticides, upon which high yielding agriculture depends, the breakdown or serious impairment of international trade and payment mechanisms could prove of serious consequence to Australia which depends on international specialisation to realise a large part of its gross domestic product. Without realisable income from its exports, without the flow of capital to finance equipment and technology, without access to the most sophisticated technology, its capacity to acquire essential imports and to keep active an economy dependent upon economic growth would be seriously impaired.

Some protection against these problems could be derived from a conscious policy of building up stocks of essential imports and expediting capital replacement programs. Time would be required to build such stocks and material, organisational and financial costs would be substantial.

The bulk of Australia's exports have in recent years gone to the same group of countries which have provided its imports, its capital and its technology i.e., Japan, EEC, and USA (Australian Bureau of Statistics, 1978/79). If their capacity to pay were impaired or frustrated by the breakdown of international financial mechanisms, Australia would face serious marketing and organisational problems for its exports of food, raw materials and fuels. While these countries, as well as many in the Third World and others not directly involved in the nuclear conflict would have need of some of these, most would lack the means of payment and barter-type agreements would be inflexible and difficult to negotiate quickly. It may well prove necessary for Australia to store important commodities over long periods in the hope that international mechanisms for exchange can be recreated. This would require Government acquisition of agricultural and pastoral products usually exported in order to sustain producers. The organisational and financial costs would be considerable. Other commodities such as minerals and fossil fuels could remain unmined without great capital loss but the consequent disruption and unemployment would pose acute problems for economic management. If imports were not available there would be less immediate need of foreign exchange earned by exports but if recovery is hoped for there would be a cumulative backlog of need.

In other words, the indirect effects alone of a nuclear conflict in the Northern Hemisphere would confront Australia with international and domestic economic and organisational problems beside which our current difficulties pale into insignificance. So long, however, as Australia and other Southern Hemisphere countries remained free from nuclear attack, it would be possible to conceive the restoration of an economy which, while it would lack the full benefit in international specialisation, could provide an adequate standard and quality of life and incorporate what is best of Western civilisation. Indeed a strong case could be made that the most important contribution Australia could make, as a participator in that civilisation or even as a member of the Western alliance, would be to preserve here the nucleus from which that civilisation could be re-established after the holocaust. Such a contribution would not be easy to achieve and it certainly could not be achieved without a major shift in the presumptions upon which Australia's foreign policies and development strategies are based.

Such a shift would be rendered more difficult by the effects of the superpower conflict on international relations generally. There will be a tendency for nations and communities to be polarised by greater or less identification (by others if not by themselves) with one or the other of the combatant groups of powers making difficult all working relations between the polarised groups. This dichotomy however will be cut across by another — the division between the aligned and the non-aligned. Horror at the consequences of the nuclear conflict is likely to induce a 'plague on both your houses' attitude around the world and a desire to evolve smaller, regionally based groupings of countries seeking simpler more self-sufficient economic systems. On the other hand that conflict might precipitate local but highly destructive wars between other powers. The social and moral divisons which these polarisations will reflect will not remain or indeed be primarily divisions between nations: they will not stop at boundaries but will divide and provoke dissension within communities, social groups and families in all societies.

In the confusion of these shifting relationships the reliability and value of even the most established and powerful alliances will become problematical. Australians will recall how during World War II the presumed support of Britain on which Australia implicitly relied suddenly was dissipated. They and others will recall how Macarthur promised the Philippines that he would return. Return he did to a land laid waste by bombing and with its society in tatters and ripe for dictatorship. So during a superpower nuclear conflict Australia's importance to its alies would be marginal. Its future would be a matter which could safely be left to be determined after the basic conflict had been resolved. Australians would be unwise to believe that 'great and powerful friends' would or could concern themselves with Australian needs or anxieties. In such circumstances Australia's important relationships may well prove to be not those with superpowers but those with the countries of its own region which will share with it the task of improvising economic institutions for survival.

Thus even without nuclear attack on Australia itself such a superpower conflict would pose economic and social problems of great complexity and would at least involve a major re-ordering of the Australian economy — a task for which we are ill-equipped by theoretical understanding or practical experience.

Direct attacks

But it is by no means certain that Australia would wholly escape such direct attacks. The evidence, so far as it is available outside the military establishment suggests that at least North West Cape, Pine Gap and Nurrungar (Woomera) are critical links in the US global strategic communications and intelligence network. The United States' most powerful deterrent force — the nuclear powered ballistic missile submarine — is dependent on a system of which the North West Cape is a critical component. Pine Gap not merely gathers up a vast volume of military intelligence but enables US to monitor Soviet missile developments and to map Soviet early warning and air defence networks. Nurrungar provides a link between the US Air Defence Command, the Strategic War Command and the Satellite early warning system — upon which US relies heavily for early warning of missile attack. (Ball, 1983, pp 9-15).

There is a variety of other installations the functions of which also are not fully known to the Australian public but it seems to be agreed that they are of lesser importance. It should be noted however that the three key installations are critical to the US strategy of deterrence, i.e., to its capacity to destroy Soviet military and industrial capacity including a large part of its population (Ball, 1983). It is hard therefore to conclude other than that at least those installations would be high in the Soviet list of priority targets.

It is even conceivable that an attack could be launched on one or several of these key installations *before* an attack on the corresponding facilities in the US as a demonstration of capability without risking the full political or strategic consequences of an attack on the US itself — a decision to make such an attack could be based on a conviction that it might be accepted by the US as not justifying a full response — in other words from an expectation that in a superpower conflict US decisions would be based solely on the strategic requirements of the conflict and that Australian lives and territory would be regarded as expendable: perhaps a plausible hypothesis.

Any consideration of economic and social effects of superpower nuclear war must therefore take account of the high probability of attack on US and joint Australia-US facilities in Australia — possibly in isolation from a general exchange of nuclear warheads between those powers.

While significant military and political opinion accepts that the three major communications centres could be the subject of attack, official policies seem to be based on the presumption that the risk of attack on other joint Australian-US facilities such as those in and near Canberra can be disregarded. Little is known by the

Australian public about the purposes or the military potential of these facilities. Soviet military analysts almost certainly know more about these matters and it is apparently believed that their knowledge will be sufficient to eliminate fear that some of them may be of tactical or strategic importance.

A much more important risk has been created by the recent decisions of the Fraser Government to increase US access to Australian naval facilities on Cockburn Sound — close to Fremantle and Perth in WA and to RAAF facilities in Darwin (decisions which so far have not been reviewed by their successors). Such access undoubtedly could be used to strengthen US defensive and aggressive capability in the Indian and Pacific Ocean regions. Whether the USSR would regard that access as justifying nuclear attack cannot be judged with certainty although the degree to which that access was in fact being used and the degree of certainty about whether weapons were carried by ships and planes using that access would presumably be important considerations. In this respect, reports that both Fremantle and Darwin, bear increasing resemblance to support centres for US military units, can scarcely lend encouragement to an expectation that the Soviet Union could feel confident on these matters.

A realistic assessment of economic and social implications of nuclear war for Australia must take into account also the uncertain but positive probability of attack on Australian facilities at Cockburn Sound and Darwin, attacks which would inevitably affect substantial urban populations and related industrial commercial and administrative facilities.

Attacks on Australia/U.S. facilities

As has been pointed out, North West Cape, Pine Gap, and Nurrungar are all located in areas of relatively low population although North West Cape is fairly close to the township of Exmouth and Pine Gap to Alice Springs. The analysts anticipate that in the event of attack casualties from blast would be confined to those actually engaged in the facilities themselves and that damage to other structures particularly those in the townships would be marginal. However, as the threat of conflict intensifies and concern develops in the related communities it may become necessary to institute elaborate civil defence measures such as planned evacuations and specially constructed public shelters as protection against prior withdrawal or panic responses even if the need for them is illusory.

The areas concerned are however in purely economic terms, of minor significance to the operation of the Australian economy. Even if analysts' expectations prove to be over-optimistic the effects of *immediate* damage i.e., from blast and fire would place no overwhelming additional burdens upon that economy. The resources required to succour the injured and those whose livelihood had been destroyed by the damage would probably be of the magnitude comparable with those which had to be called upon in natural disasters like the Darwin cyclone and the recent bush fires in South Australia and Victoria.

The possibilities of casualties from radiation fallout as a result of even these limited attacks are more serious, especially for South Australia where, if meteorological conditions were unfavourable, there could be extensive casualties in provincial cities and significant damage to continuing health in Adelaide particularly in the form of reduced resistance to other diseases. This area is an important source of food and other agricultural products both for city consumption and for export. Not enough information is available to support a judgement about whether fallout would render some significant part of this production unusable or whether the soil itself would be seriously contaminated (Ball, 1983).

The major economic impact would show in the provincial cities where medical care facilities would be seriously strained and manpower depleted. The contribution of these cities to the Gross Domestic Product would for a period of months and possibly years, be drastically reduced. If, however the rest of the economy were not directly affected there would be sufficient resources available from there for these problems to be dealt with in emergency fashion and recovery initiated as has been possible from natural but local disasters such as bush fires or cyclones.

If the populations of these regional cities, like Whyalla, Port Augusta and Port Pirie as well as Adelaide were aware of the risks to which attacks on facilities such substantial distance from them rendered them liable, Governments and other authorities would probably be pressed to establish civil defence and medical care facilities in advance if not to change other more fundamental aspects of their policies. Only evacuation would seem to offer significant protection against fallout to the populations involved —

a measure difficult to plan and to organise in circumstances where warning times would be short, the effect of meteorological conditions unpredictable and where civilian fears would be exaggerated by ignorance and distrust.

However one must conclude that the consequences of nuclear attacks confined to Australia-US facilities at North West Cape, Pine Gap and Nurrungar would be within the capacity of a healthy economic and political society to deal with. There would probably be serious political dangers. Historical evidence suggests that disasters generally produce a unifying response in the population of a community as a whole — especially if the disaster is locally confined, and within the capacity of the goodwill, resources and generosity of the rest of the community to handle. This certainly seems to be true of natural disasters which can be attributed to 'acts of God' and for which no human blame can be laid. There seems some reason to believe that this would be less true of nuclear attacks where the grossly altered appearance of persons who had suffered severe burns and the fear of future casualties would evoke an intense emotional impact (F. C. Ikle, 1958, pp.27-34). It would be strange however if disasters which clearly would be 'man made' did not provoke significant hostility: if there were not a desire to look for a scapegoat. So long as this hostility was directed solely at the foreign power which proximately precipitated the disaster, its domestic social effects, at least in the short run, would not be likely to impair that unifying response. It would however be surprising if some of that blame did not fall on those closer at hand especially as the disaster would be directly associated with and could be seen as caused by decisions of the existing and past governments on foreign policy and defence.

Those who recall the controversies in Australia during World War II over the proposed 'Brisbane line' and the suggestion that all Australia outside the 'heartland' containing Newcastle, Sydney, and Wollongong, might be abandoned to Japanese invaders will probably doubt whether those affected in the South Australian cities would be greatly impressed by the strategic analyses and political judgements which exposed them to its effects.

Attacks on Darwin and Cockburn Sound

Much more serious however would be the effects of attacks on

Cockburn Sound or on Darwin. Analysts believe that blast from a weapon detonated over the establishment in Cockburn Sound would be sufficient only to cause the collapse of houses and to start small fires over residential areas up to the outskirts of Fremantle itself. However nuclear fallout would deposit radiation over an extended area of Fremantle and Perth with probable fatalities of the order of 100 000 people and a corresponding number of casualties from radiation sickness.

While the Darwin population is much smaller, destruction would be more absolute. Blast from a nuclear bomb exploded above the RAAF base would produce fatalities involving half the total population with injuries affecting an additional 40 per cent and with much of Darwin probably being covered with lethal levels of fallout (Ball, 1983, pp.42-3).

Such catastrophes have social and economic dimensions difficult to conceive let alone define. Apart from the large number of deaths, injuries, damage to food and productive land in the region, and the destruction of all the existing mechanisms of economic intercourse there would be overwhelming demands on hospitals, medical services, homecare and family support in circumstances where the material and professional and human resources would, in Darwin at least, have been almost obliterated. Added to these effects there would be psychological impacts arising from fear of further attacks and other anxieties at least as significant. Competition for medical supplies and other forms of immediate aid could set victim against victim and intensify the suppressed racial, class and economic hostilities that exist even in the healthiest society (Katz, 1982, p.68).

But let us be clear. Ghastly as these prospects are they would not in themselves mean the destruction of economic life for Australia. Perth and Darwin, however important to those who live in or otherwise identify with either of them, are not crucial to the survival of Australia in a narrowly economic sense though their destruction would perhaps threaten social stability and cohesion more seriously than attacks on the main communications bases even in a long term political sense. When I visited Hiroshima with Ben Chifley in 1946 I thought the destruction wreaked there marked the end of human life in that area. Yet within a decade Hiroshima was rebuilt and again a functioning modern city. Of course the identity of the name obscured the facts that the city after that decade housed an almost wholly new population and productive enterprises newly established and that the few survivors of the original Hiroshima population were physical and psychological cripples living in perpetual fear and anxiety, viewed by their contemporaries with a strange mixture of pity, suspicion and hostility.

The economic system, as distinct from those who live by it has great resilience. So long as its basic fabric is maintained it is capable of mobilising human and technological resources to restore its fragmented components to effective operation. But the maintenance of that basic fabric requires first that a substantial part of the total population and the human skills and experience it comprehends survives and second that the institutions which underlie the economic system continue to function however imperfectly. We tend to forget that the economic system takes for granted the social institutions of property and its protection by the law and its agents; the existence of currency which is legal tender and a financial system for the care and protection of wealth expressed in its terms; the complex system of contracts and the means for their enforcement, structures like corporations with perpetual succession capable of owning property, suing and being sued, and laws which grant to their proprietors the protection of limited liability and ultimately if necessary of resort to bankruptcy. This whole framework, without which the contemporary economic system would quickly disintegrate is fragile indeed, depending as it does on the acquiescence — indeed the subordination to its preservation of the self-interest of those who constitute the community. Uncertainty about the permanence of this framework is quick to end that subordination. Looting and pillage are endemic in all major disasters. When society ceases to protect the individual it is each for himself and the devil take the hindmost. The experience of Darwin in 1942 is instructive (Hall, 1980).

Fremantle, Perth and Darwin after attack would almost certainly suffer from some, probably most, of these effects. If they were to survive them and recover as cities it would as in the case of Hiroshima, be the result of effort initiated, and resources supplied, from outside.

Attacks on cities and industrial complexes

It is therefore disasters which threaten to destroy a large proportion of the population of a society or which make its social institutions unworkable that threaten the capacity to survive the disaster and make it impossible for the productive system of that society to initiate and sustain recovery programs. The possibility therefore of destruction in Australia of the order envisaged in the Northern Hemisphere derives from the risk that attacks might be launched on industrial complexes and cities. Defence analysts believe that the Soviet does not nor could readily possess warheads in sufficient number to mount such attacks without endangering its capacity to target its primary enemy or enemies effectively. They therefore regard a Soviet decision to use them for this purpose as highly improbable and believe that the risk can be in effect, ignored.

It is probably reasonable to accept that, in terms of cost-benefit analysis a scenario incorporating such a decision is implausible. But in conflicts of the kind envisaged between the superpowers decisions may not be made in the light of such analysis. It is possible to conceive of circumstances where the need to conserve warheads was no longer felt — e.g., at the point of certain defeat — where a decision like that of Samson to pull down the pillars of the temple might form part of a final suicidal gesture. Apart from such dramatic scenarios, policy must take account, not merely of the assessed likelihood of a disaster, but its probable magnitude if, however improbable, it in fact occurs. A gambler may light-heartedly bet on a five-to-one chance if what he puts at risk is small in relation to his total resources. But he is unlikely to stake his all even on what seems a near certainty. Nor is he likely to take the five-to-one chance of Russian roulette where his life is at risk. Unfortunately the decision about whether the risk is to be accepted will be made by those who alone will have some chance of ensuring their own personal survival.

To gamble with the lives of millions on a judgement about how desperate men would behave in a time of crisis and despair is, to say the least, not rational. Even outside such a context there is little warrant for confidence. The first atomic bombs were dropped — not by a beaten foe but by a power within sight of inevitable victory — in circumstances where, some historians have argued, those responsible for the decision did not expect it significantly to shorten the war.

The possibility of attacks on major population centres and industrial complexes cannot therefore be wholly disregarded. However the scope and effects of such attacks have not, at least in the public context, been closely analysed because of the inherent difficulty of the task but also because the contingency seems to those responsible too implausible to warrant attention. However an assumed attack on Sydney would result in immediate fatalities from blast and heat of 180 000; fatalities from fallout would number about 480 000; there would be about 350 000 injured people; but the majority of the population would not be directly affected. Similar results could be effected from an attack on Melbourne (Ball, 1983, pp.43-4).

The purpose of such attacks would be to incapacitate the economy and so render impossible active participation in the war or support for allies so engaged. This would be achieved by destruction of the working population, and of managerial and professional skills, by damage to key industrial plants, transport equipment, and sources of fuel and power. The effect of such destruction and damage, even if they were not wholly comprehensive would produce bottlenecks in the productive processes and so spread the effects of immediate damage through a major part of the economy. There is little doubt this result would be achieved by attacks on major Australian cities and industrial complexes.

The possibility of limiting the human casualties by civil defence measures would rest on the length of warning times. If attacks came in isolation from more general hostilities these times would be measured in minutes — and would provide no opportunity for effective protective action. However analysts believe that nuclear attacks are most likely to emerge from escalating conflict conducted initially, perhaps, by non-nuclear weapons but progressing through the use of smaller nuclear tactical weapons to major exchanges of megaton weapons. Consequently they judge that there would be periods of strategic warning times sufficient to provide time to plan and develop civil defence responses. Such responses would rely essentially on combinations of evacuation and public and private shelters against blast and nuclear fallout. If effectively planned and organised they could, it is believed, reduce the loss of life and other casualties but the measures would present great organizational difficulties and, even if these were overcome, loss of life and casualties would remain horrendous.

Even if strategic warning times allowed evacuation plans to be developed the problem of at what point to put them into effect would remain. The time between the despatch of the missiles and their arrival would still be too short for action. Decision to evacuate must therefore be taken in anticipation and the risk of error would remain. Unnecessary evacuation especially if repeated would damage the credibility of the civil defence authorities and of course a decision to evacuate taken too late would be a disaster.

Briefly the prospect of attacks on Australian cities and industrial complexes during a Northern Hemisphere nuclear conflict, while probably unlikely, cannot be ignored; if such attacks were mounted civil defence measures could limit loss of life but would be most difficult to organise up to the point of almost certain breakdown according to individual circumstances; even with such measures the devastation and human casualties are likely to be comparable with those on strategically more important targets in the Northern Hemisphere.

In the absence of detailed studies one can do little more than attempt to apply the outcome of such studies in other countries such as USA to Australia so far as the circumstances are comparable. Australia is highly urbanised and has few major industrial and population centres. Although some of those centres are widely separated single attacks on a relatively small number of them could for the period of hostilities incapacitate the Australian economy for all but disaster oriented activities and render it militarily largely ineffective.

The whole urban infrastructure would certainly be chaotic and seriously damaged — including transport, power, housing, fuel and food distribution, and sanitation. Its restoration would certainly be a priority demand on resources and the attention of planning authorities. Apart from the actual loss of workers from death, injury and illness, fear of further attacks may well make it difficult to persuade survivors to return to their places of work. The loss of skilled workers, supervisors, managers and the operators of computers and like equipment upon which economic activity increasingly depends would make extremely difficult even the partial recovery for which quantitatively human and material resources remained sufficient. Especially disruptive would be the probable loss of financial records relating to both persons and enterprises.

During World War II when Japanese invasion seemed imminent and the possibility that parts of Australia would be cut-off and isolated I was involved in the setting up of emergency supplies of essential goods at appropriate regional locations to ensure at least temporary survival of the beleagured population. The task was made especially difficult by the absence of any traditionally based regional administrative and executive mechanisms. The realization of this deficiency was one of the factors which, in the post-war period, stimulated a lively interest in regional planning. Today the risks of such isolation are much greater and the centralisation at both State and Federal levels even more intense. The task of improvising decision making and administrative processes in areas not seriously damaged but isolated would add significantly to the problems of nuclear attack — especially if such areas had become the recipients of the evacuated population — perhaps many times the local numbers.

These effects would be superimposed on the problems created by the isolation of Australia from its sources of imports and external capital, technology and managerial capacity to which I referred earlier.

They would too have to be faced in a context of devastating psychological and emotional effects on a population which has no experience of war within its own territory. Abnormal behaviour often anti-social in its implications is likely to reflect the general disorientation — fear, doubt, apathy, hostility towards those in authority and the intensification of antagonisms normally suppressed are to be expected (Katz, 1982, p.67).

In the event of attacks on cities and industrial complexes the description of the US economic system quoted in the first part of this paper will apply generally to Australia — i.e., 'the economic system will be partitioned and fragmented; its integration as an effective system seriously in question'. It is impossible to judge whether the disruption would be such as to preclude recovery in the longer term. So much would depend on whether effective smaller components of the system remained substantially undamaged, whether help could come from outside the country and whether the social, administrative and political structures survived sufficiently to function. It is these structures which would be the most threatened elements of civilisation (Ikle, 1958, p.229).

The possibility that there would be a prolonged period of social chaos comparable with the aftermath of the Black Death in Medieval Europe or the period of Civil War after the Revolution in Russia in 1917, cannot be rejected.

I have tried in this paper to consider the implications of nuclear war for Australia, within the framework of policy decision making from an essentially economic point of view. I have however been increasingly conscious how unreal such an approach is. The questions raised by the possibility of nuclear conflict must be answered by each of us, whatever our speciality, as whole persons not as a representative of a discipline or an institution of limited purpose. And as whole persons we will or should be guided by the values we hold: values to which nuclear war is utterly abhorrent.

References

Australian Bureau of Statistics (1982) *Imports Australia* (1978/79) *Overseas Trade — Australia.*

Australian Parliament, Joint Committee on Foreign Affairs and Defence (1981). 'Threats to Australia's security: their nature and probability'. AGPS, Canberra

Ball, D., (1980), *A Suitable Piece of Real Estate: American Installations in Australia,* Hale & Iremonger, Sydney.

Ball, D. (1983), 'Limiting damage from nuclear attack'. In Ball, D. and Langtry, J. O., *Civil defence and Australia's security in the nuclear age,* Strategic and Defence Studies Centre, ANU, Canberra.

Caldicott, H. (1978), *Nuclear Madness,* Jacaranda Press, Milton, Queensland.

Dahlitz, J. (1983), *Nuclear arms control and effective international agreements,* McPhee Gribble, Melbourne.

Hall, T. (1980), *Darwin 1942,* Methuen Australia, Sydney.

Hinchcliff, J. (ed.) (1981), *Confronting the nuclear age — Australian responses,* Pacific Peacemaker.

Ikle, F. (1958), *The social impact of bomb destruction,* University of Oklahoma Press, Norman, Oklahoma.

Katz, A. M. (1982), *Life after nuclear war: the economic and social impacts of nuclear attacks on the United States,* Ballinger Press, Cambridge, Massachusetts.

Mathams, R. H. (1978), 'Strategic weapons and their effects' in *Report on Civil Defence Study conducted at Canberra ACT 20-22 March 1978,* (Department of Defence, Annex B to DES 92/1/20), pp.3-4.

Richelson, J. (In Press), 'Strategic nuclear targeting', In Ball, D. and Richelson, J., eds., *Assessment of the post-attack environment,* Chapter 13.

8 THE ATMOSPHERIC EFFECTS OF NUCLEAR WAR

A. Barrie Pittock

A. Barrie Pittock

Do Atmospheric Effects Matter?

The US National Security Council estimates that 140 million Americans and 113 million Soviet citizens would die from the immediate effects of a major nuclear 'exchange'. So why should we bother to inquire into the effects on the atmosphere? A conventional 'scientific' answer is -'because it is an interesting problem'. But that is an inadequate ânswer. In science there is an infinity of questions which can be asked, but we have neither the time nor resources to answer them all. Ultimately scientists and their employers ask and seek answers to questions because they have some social or economic purpose, even if in so-called 'basic' or 'pure' research that purpose is cultural or otherwise indirect.

We need to ask and answer questions about the environmental consequences of nuclear war because they bear on some fundamental social questions of our time: Can nuclear war serve any rational purpose, even for the 'winning' side? Can a nuclear war be 'won'? Will there be any long-term survivors? Would the unintended 'side effects' be as bad or worse than the intended military effects? Is nuclear war 'worth it'?

For many in the peace movements these questions are unnecessary, because for them the mass killing of human beings, which is inevitable in nuclear war, can never be justified. However, there are many people who believe that nuclear war, or at least the threat and risk of it, may be justified in the 'defence' of certain economic, social or ideological causes. The question of the environmental consequences are important since even these people may become convinced that nuclear war is not a rational option if the consequences can be shown to be bad enough.

The Growth of Interest in the Problem

Early interest in the problem of the global scale effects of nuclear weapons concentrated on radioactive fallout and led in 1963 to the banning of nuclear weapons testing in the atmosphere. In 1966 Batten addressed the question of the effects of nuclear war on the weather and climate in a Rand Corporation report for the US Atomic Energy Commission. Batten stressed possible effects of debris injected into the upper and lower atmosphere on cloud formation, precipitation, and the solar and heat radiation entering and leaving the atmosphere. In addition he suggested that extensive fires might change the surface characteristics and modify local weather patterns. He concluded that 'the interference with the atmospheric processes in some cases can be sufficient to produce changes in them; however, the nature, extent, and magnitude of the resulting anomalies in the weather and climate are uncertain'.

The United States National Academy of Sciences in 1975 produced a report entitled 'Long-Term Worldwide Effects of Multiple Nuclear-Weapons Detonations'. Based on a worst case scenario in which 10 000 million tons of TNT equivalent are exploded, half of this in warheads large enough to place debris into the upper atmosphere, the Academy Committee concluded that 'the major predicted long-term effects are thought to derive not from the dispersion of radionuclides but from an increase in the flux of ultraviolet radiation on the earth's surface due to reduction in the stratospheric concentration of ozone'. The Academy President in his letter of transmittal cautioned that the Academy committee 'deliberately refrained from synthesizing an integrated vision of this catastrophe' such as 'the social, political, or economic consequence to the rest of the highly interdependent world civilization of the hypothesised nuclear exchange'. He also cautioned readers about possible climatic instabilities.

The Academy committee's focus on the ozone reduction problem was due to the recent emergence of this problem in relation to stratospheric injections of oxides of nitrogen from supersonic aircraft emissions (Crutzen, 1970; Johnston, 1971). The much less serious aircraft emission problem was intensively studied in the Climatic Impact Assessment Program (Grobecker, Coroniti and Cannon, 1974).

A review of the human and environmental effects of nuclear weapons, by a worker at the Los Angeles weapons laboratory, was published by Mark (1976). The standard work on the effects of individual nuclear weapons was first published in 1957, and in revised form in 1962 and again in 1977 (Glasstone and Dolan). Neither of these publications gives much attention to global scale effects on the atmosphere but they provide useful background. Such effects are briefly dealt with, however, in a study by the US Office of Technology Assessment (1980) and by Westing (1977, 1981). Significantly, the OTA gave as its number one finding: 'The effects of a nuclear war that cannot be calculated are at least as important as those for which calculations are attempted. Moreover, even these limited calculations are subject to very large uncertainties'.

In 1982 Jonathan Schell, in *The Fate of the Earth,* argued that the threat to the global ecosystem from a possible nuclear war is such that the survival of human life on earth is in question. While not claiming any certainty, he cited the possible synergistic effects of radioactive fallout, ultraviolet radiation enhancement due to ozone depletion, epidemics, widespread destruction of medical supplies, contamination of food and water supplies, etc., as leading to the possible extinction of the human race as the survivors succumb to the degraded environment. Importantly however, Schell did not specifically discuss effects in the Southern Hemisphere of a nuclear war fought essentially in the Northern Hemisphere.

The editors of *Ambio* in 1980 sponsored an international project to draw up a scientific appraisal of the human and ecological consequences of nuclear war. This was published in 1982. Notable in the context of the present review is the paper by Crutzen and Birks (1982) entitled 'The Atmosphere After a Nuclear War:

Twilight at Noon'. This paper laid great emphasis on the possible effects of the input of smoke and dust into the lower atmosphere in reducing the intensity of incoming solar radiation in parts of the Northern Hemisphere to levels below that required for photosynthesis for a period of several weeks. The authors suggested that this might be followed by a widespread photochemical smog with potentially damaging concentrations of ozone and other photochemical products.

On the basis of the plausible nuclear war scenario proposed by the *Ambio* advisors as possible in 1985, in which far fewer large nuclear warheads (i.e., greater than 500 kilotonnes of TNT equivalent) would be exploded than in the NAS (1975) scenario, Crutzen and Birks considered that drastic reduction of the protective layer of ozone in the upper atmosphere was unlikely. They did however reconsider the effects on the ozone layer of the earlier NAS scenario and reached similar conclusions to the NAS report, viz. reductions of the total ozone column in mid northern latitudes ranging from 40 to 65 per cent up to two years after such a war, and 10-20 per cent maximum reductions in the Southern Hemisphere. There were many uncertainties in Crutzen and Birks's calculations, and several major revisions will be mentioned below and in the accompanying paper by Galbally, Crutzen and Rodhe (this volume).

Two American research groups with wide backgrounds in atmospheric optics and the study of planetary atmospheres, and who have studied the possible effects of the impact on Earth of a comet, asteroid or large meteor (Pollack, Toon, Ackerman and Mackay, 1983; Turco, Toon, Pollack and Sagan, 1981; Gerstl and Zardecki, 1982), have recently turned their attention to the somewhat analogous problem of the effects of a major nuclear war. Interest in the impact problem arises from the hypothesis that a collision of a large body from outer space with Earth may explain the widespread extinctions of dinosaurs and other reptiles about 65 million years ago.

Gerstl and Zardecki (1983) suggest that the figure given by Crutzen and Birks (1982) of an equivalent global atmospheric dust and smoke amount of about 10^{15}g, (i.e. 1000 million tonnes) after a nuclear war is about 10 per cent of the estimated minimum required stratospheric dust load necessary to reduce photosynthesis to one thousandth of normal. They comment that, as 'smoke particles are at least a factor of 10 more absorbing than the dust' considered in their analysis of an asteroid or comet collision with Earth, Crutzen and Birks's figures 'would imply also a severe darkness scenario'.

Turco et al. (1982) report:

> We have performed a variety of sensitivity studies to define a range of possible outcomes of a full-scale nuclear exchange. In some cases we predict long-term effects which are small in comparison to the primary nuclear destruction due to blast, thermal pulse and local radioactive fallout. However, a significant number of cases show potentially devastating global effects. In these circumstances, a combination of stress caused by severe climatic perturbations (surface cooling of 10°C or more), radiation doses in the tens of rem, and tenfold increases in UV-B solar radiation exposures, together with widespread shortages of food and potable water, epidemics, serious injuries and lack of medical facilities and supplies, cumulatively imply widespread death in man (sic) and possible extinction of numerous land and marine species.

The ecologist Paul Ehrlich has attempted to draw together the whole gamut of ecological effects of a major nuclear war, including atmospheric effects. 'In short, the ecological effects alone of a large-scale nuclear war that occurred in the spring, summer or fall could destroy civilization in the Northern Hemisphere. Those of a winter war, when much of the combatant nation was under snow, almost certainly would be smaller but still catastrophic.' (Ehrlich, in press).

There have been an increasing number of national and international studies and symposia on the general effects of nuclear war. To date very few of these studies have explicitly considered the question of the environmental impact on the Southern Hemisphere. In my earlier review, largely based on the Crutzen and Birks paper, (Pittock, 1982), I made an attempt to examine the Southern Hemisphere impact. My conclusion then was that human survival might be threatened in the Northern Hemisphere, but that human beings would probably survive in the Southern Hemisphere, notably in those areas of South America and Africa not likely to be directly involved. New considerations outlined below, notably a possible increase in inter-hemispheric transport and the convective mixing upwards of dense smoke and dust clouds into the stratosphere, now lead to increasing concern at the possibility of major climatic disruption in the Southern Hemisphere in the event of a full-scale nuclear war. Nevertheless it remains true that some humans would probably survive in the Southern Hemisphere. Due to the many uncertainties and complexities involved, however, this is not a firm or confident conclusion.

A Scientific Setting

(i) *Atmospheric Transport and Residence Times*

The basic temperature structure of the unperturbed atmosphere is shown, in a north-south vertical cross-section, in Figure 1.

The lower atmosphere where temperature is decreasing with altitude is called the 'troposphere', and is a relatively well-mixed region where convective overturning occurs, with cloud formation and precipitation. Particulate and water-soluble debris or combustion products in this region, i.e., below the upper boundary of the troposphere (the 'tropopause') is normally removed in several days to a week. Above the tropopause the temperature increases with altitude up to heights of about 50 km. This region, termed the 'stratosphere', is much more stable, with few clouds and almost no precipitation. Gaseous contaminants, and particles too small to fall out rapidly under the influence of gravity, will in general stay in the stratosphere, once introduced, for periods ranging from six months to several years depending on altitude, latitude and season.

Exchange of air, and of contaminants therein, between the Northern and Southern Hemispheres normally occurs in one or two years. Thus, except in some local tropical monsoonal circulations, major transport of atmospheric effects of a nuclear war from the Northern to the Southern Hemisphere will occur in the stratosphere (unless there is a major departure from normal circulation patterns). Tropospheric contaminants will tend to be removed before they have time to be transported into the Southern Hemisphere. We would thus expect that, if the atmospheric circulation remained unchanged, the major contribution to contamination of the Southern Hemisphere troposphere would be from bombs detonated in this hemisphere. According to the *Ambio* scenario (*Ambio* advisors, 1982) only some 3 per cent of the total megatonnage is likely to be exploded in the Southern Hemisphere, so in this respect our chances of survival are much better than in the north.

Figure 1

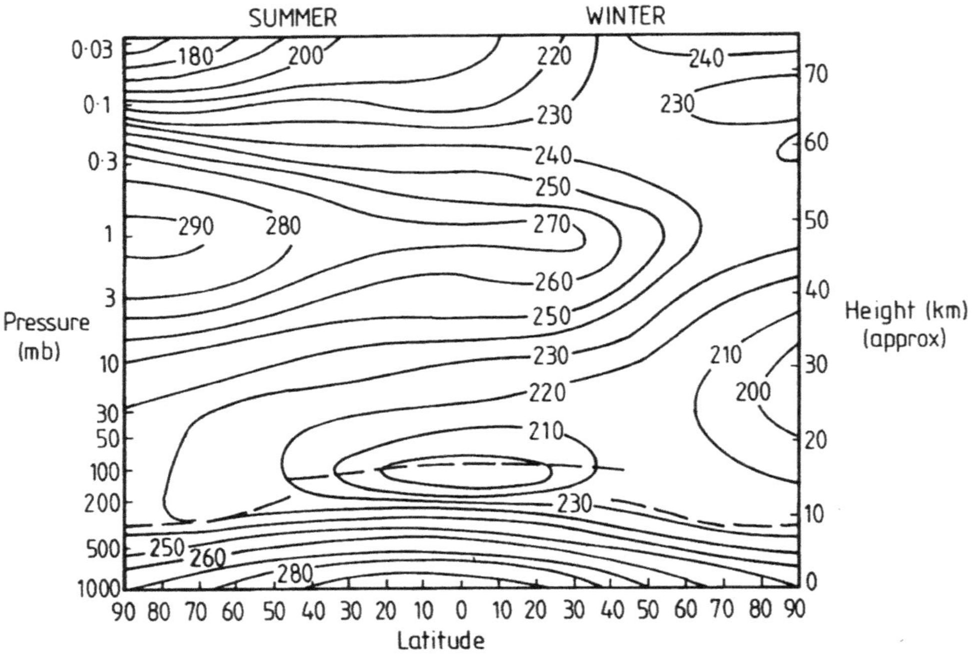

Average north-south vertical cross-section of atmospheric temperature in degrees Kelvin (°C plus 273). The dashed lines indicate average location of the tropopause, or upper boundary of the well-mixed lower atmosphere (the troposphere). The region above the tropopause, known as the stratosphere, is more stably stratified

However, contaminants which are injected into the Northern Hemisphere stratosphere, because of their much longer residence time in the atmosphere, are more likely to be transported to the Southern Hemisphere. This would be the most likely cause of widespread major atmospheric effects in the Australian region, apart from direct attacks on our major cities. A critical question then, apart from possible major departures from normal atmospheric circulation patterns (which will be discussed below), is how much debris and contaminants will be injected into the Northern Hemisphere stratosphere in a major nuclear war.

(ii) *Injection into the Stratosphere*

There are at least four possible major sources of contaminants into the stratosphere from a nuclear war. One is from detonations in or above the stratosphere, e.g., from nuclear armed anti-ballistic missiles, or high altitude explosions designed to damage electrical circuits via the electro-magnetic pulse effect (Hutton, 1982; Broad, 1981). Such detonations would introduce oxides of nitrogen and fission radionuclides. A second source is from nuclear explosions in the troposphere powerful enough that their fireballs penetrate the stratosphere due to the energy of the explosions themselves. Figure 2 is a general representation of the relationship between the height of the fireball cloud tops and bottoms and the weapon yield. This varies with latitude because of the differing vertical stability of the atmosphere with latitude. With yields greater than 400 kilotons more than half the cloud penetrates the stratosphere at latitudes greater than about 30 degrees, while nearer the equator a yield greater than one megaton is needed.

The third source of atmospheric contaminants is from major urban fires initiated by nuclear explosions which by themselves would not be energetic enough for their bomb clouds to penetrate the stratosphere. According to a study by P. C. Manins, which will be published elsewhere, the majority of large cities (approximately 200 out of 284 with populations greater than one million people) would give injections of debris which would just reach the stratosphere at latitudes polewards of 30 degrees. Some large, high-density cities would give injections to 19 km altitude, well into the stratosphere at these latitudes, while the burning of many medium-density cities would inject material to heights of 10 to 16 km. Manins concludes that fires from other sources such as forests or oil fields would be unlikely to inject material into the stratosphere,

Figure 2

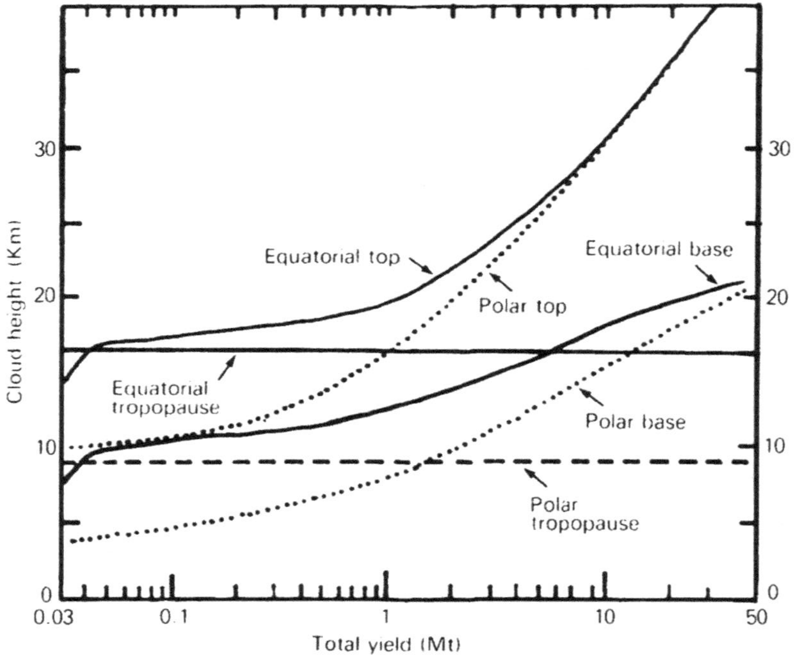

Top and base of the nuclear explosion mushroom cloud as a function of yield of the bomb in millions of tonnes of TNT equivalent. Average heights of the equatorial and polar tropopause are indicated. The bomb clouds rise further in the equatorial atmosphere because it is in general less stable than the polar atmosphere. After Peterson (1970)

although they would be significant sources of pollution in the troposphere.

A fourth possible source of stratospheric contaminants is the convective mixing upwards of dense smoke and dust clouds, initially injected into the troposphere, due to rapid heating of the upper cloud layers by the absorption of incoming solar radiation. Preliminary calculations, and model experiments in a tank of stably stratified water, suggest that optically dense clouds may warm sufficiently to mix upwards against quite strong temperature gradients.

Rising fireballs from large nuclear explosions will inject into the stratosphere not only oxides of nitrogen, generated by the high temperatures in the shock waves, but also fission radionuclides and surface material entrained in the fireball in the form of dust and vaporised solids, only some of which will condense, coagulate and fall out in the first few days. Major urban fire plumes will inject similar material plus much larger quantities of sooty material and ash, much of it highly absorbing of incoming solar radiation.

(iii) *Atmospheric Chemistry*

Several major sets of atmospheric chemistry processes are relevant to the problem of the impact of nuclear war. One is the tropospheric set of reactions best known in relation to photochemical smog situations (Calvert, 1982). Chief ingredients in these tropospheric reactions are oxides of nitrogen, both generated by the nuclear explosions themselves and by subsequent fires, and hydrocarbons produced by incomplete combustion in urban, rural and oil or gas fires, or released by the evaporation of oil spills and natural gas leaks. These will react in the presence of sunlight to produce abnormally high concentrations of ozone and other obnoxious chemicals which are damaging to delicate plants and sensitive animal tissues such as eyes, nose, throat and lungs. Formation of these harmful products in surface air may be at least initially suppressed by the reduction of available sunlight due to dust and smoke clouds in the troposphere and stratosphere. The duration of this reduction of available sunlight is an important issue.

A second major set of chemical reactions operate in the stratosphere to control the concentration of ozone (Crutzen, 1979; National Research Council, 1982a) which is normally present in much higher concentrations in the middle stratosphere than in the troposphere. The amount of ozone in a vertical column largely controls the intensity of solar ultraviolet radiation in the biologically damaging wavelengths (known as UV-B radiation) which produce sunburn, skin cancers and damage to the cornea of the eye (leading to cataracts and blindness). At the temperatures and ultraviolet radiation levels which prevail in the stratosphere additional oxides of nitrogen lead to a reduction in ozone concentration. The introduction of bomb- or combustion-generated oxides of nitrogen into the stratosphere will lead to reductions in the ozone column amounts and to increases in UV-B intensities at the surface. Again, this effect at the surface would initially be offset by the presence of absorbing dust and smoke layers in the troposphere and/or stratosphere so that the lifetime of these absorbing layers is critical.

The third major chemical consideration is the process by which fine particles are generated *in situ* by gaseous contaminants. This process operates naturally after major volcanic eruptions such as the El Chicon eruption in Mexico in April 1982. This process leads to the continuing formation of small particles, replacing those lost by coagulation and subsequent gravitational fallout, and could be important in prolonging the lifetime of absorbing layers in the stratosphere.

(iv) *Climatic Effects*

Possible climatic effects could arise from many different mechanisms, the obvious ones being modification of the incoming solar radiation by dust, smoke and other contaminants in the atmosphere, and changes in the characteristics of the earth's surface due to fire, deposition of soot, and the killing off of large areas of vegetation.

The nature and seriousness of the dust and smoke layer problem depends on how much is injected into the stratosphere where it would have a lifetime of years as opposed to days or weeks in the troposphere. The optical properties of the particles are also important since non-absorbing particles modify the incoming solar radiation only by scattering. That portion of the solar radiation which is scattered in a forward direction may still reach the surface, but that which is scattered back into space is lost to the earth's heat budget. Large numbers of particles, however, lead to multiple scattering which further reduces the proportion of incoming radiation reaching the surface. Absorbing particles remove a proportion of the incoming radiation so that its energy heats up the absorbing layer instead of the surface of the earth. Consideration must also be given to the absorbing properties of the particles at infra-red wavelengths at which heat is radiated to space from the earth's surface.

Absorbing particles in the stratosphere cause local heating in the stratosphere and cooling at the earth's surface (see e.g., Lenoble, Tanre, Deschamps and Herman, 1982). For example, the eruption of Mt Agung volcano in Bali in 1963 caused a warming of several °C in the stratosphere, and an estimated surface cooling of a few tenths of a degree (Hansen, Wang and Lacis, 1978). If a large number of urban fire plumes penetrate the stratosphere, or if tropospheric smoke clouds convectively rise into the stratosphere as discussed in section (ii) above, then given the high absorbtivity of soot particles, surface cooling of the order of several degrees or more would be possible on a time-scale of years. Shorter lived but much more severe surface cooling over land might also be expected from the high concentration of dust and smoke injected into the troposphere (see e.g., Coakley, Cess and Yurevich, 1983). This is the basis of the Turco et al. (1982) estimate of a 10°C or more cooling, which would be disastrous for many crops, natural vegetation, humans and animal life, depending upon the season and the duration of the cooling. The average global surface temperature difference between the present and the depths of the last ice age some 18 000 years ago is less than 10°C.

Ocean surface temperatures would be less rapidly cooled by dust and smoke in the atmosphere because of the great heat storage capacity of the upper mixed layer of the oceans. This would set up unprecedentedly sharp temperature gradients between cold land masses and relatively warm oceans, which would result in air subsidence and outflow over the continents, with reduced rainfall over land, and the reverse situation over the oceans. Overall rainfall would however be reduced due to cooler air temperatures and a progressive reduction in total evaporation as the ocean surface gradually cooled. This effect would be marginally increased by a reduction in land-surface evapotranspiration due to the killing off of large areas of forest and other vegetation by fire, radioactivity, surface cooling, etc. (Shukla and Mintz, 1982).

Complex feedbacks and chains of cause and effect would follow, leading to unpredictable local weather and climatic anomalies. Changed surface winds would alter the wind-driven ocean circulation. Local atmospheric cooling also leads to a displacement of surface ocean currents by changing the horizontal density gradients in the surface layers of the ocean (Nof, 1983). These changes in ocean currents would in turn modify sea surface temperature distributions and the effect of the ocean on the atmosphere.

Another important effect could be that the initial large input of dust and smoke into the Northern Hemisphere troposphere would set up an unprecedented surface temperature gradient between the Northern and Southern Hemispheres, which might lead to enhanced low level flow of air into the Southern Hemisphere similar to an enhanced Australian summer monsoon (which already involves some cross-equatorial flow in the Indonesian region). Such enhanced effects have not been modelled, but they could be effective in transporting pollutants into the Southern Hemisphere.

(v) *Particulate Coagulation*

One of the principal processes operating in clouds of dust or smoke is the coagulation of small particles into larger ones which, at radii greater than 0.25 μm, are less effective per unit mass in scattering and absorbing light, and which will fall out faster due to gravitational settling. This coagulation process is due to 'Brownian motion', in which collisions with air molecules move the dust or smoke particles around in random fashion, causing them to collide. The rate of coagulation varies roughly as the square of the concentration of particles. In surface air very high concentrations of particles coagulate within a few days into concentrations no greater than about 10 000 per cubic centimetre (Twomey, 1977) unless new small particles are being injected or formed to replace those lost by coagulation. At higher altitudes lower air pressures and temperature lead to slower coagulation rates, and these also depend on the surface properties of the particles which determine the efficiency with which they stick together.

Crutzen and Birks (1982) in their calculations on the sunlight absorption due to tropospheric dust and smoke neglected coagulation, but they postulated a continuing input of new particles due to fires burning for weeks on end. Pollack et al. (1983) and Gerstl and Zardecki (1983) did allow for coagulation in their calculations on the effect of dust clouds from collisions between Earth and various cosmic bodies. In this latter case coagulation quickly reduces much higher particle concentrations to concentrations within one or two orders of magnitude of the nuclear war situation. Given that smoke particles are far more highly absorbing than the ejecta material from a cosmic collision, Gerstl and Zardecki (1983) conclude that comparable effects might be anticipated. If greater quantities of dust and smoke are injected into the stratosphere than was contemplated in Crutzen and Birks (1982), the analogy with a cosmic collision may be even closer.

(vi) *The Experience of the Atmospheric Test Series*

In relation to the global distribution of radioactive fallout and what it might tell us about the transport of other atmospheric contaminants, the series of 370 atmospheric tests of nuclear weapons over the period 1945-1975 (Carter and Moghissi, 1977; Enting, 1982) is a useful indication of what might be expected in a major nuclear war.

Enting (1982) gives an estimated total yield from the tests of about 500 megatons of TNT equivalent, of which 90 per cent was from 90 detonations of 1 megaton or greater for which most of the fission products would have entered the stratosphere. In comparison, the *Ambio* scenario (*Ambio* advisors, 1982) envisages a total yield in a major nuclear war of 5742 megatons. The majority of this yield is due to surface bursts and relatively small warheads which would inject relatively small amounts of radioactive debris into the stratosphere. According to calculations done by Enting for the symposium (to be published elsewhere), the 'effective yield' as far as stratospheric input is concerned would be only some 1482 megatons, or about three times the total from the atmospheric test series.

As far as long-lived fallout is concerned, such as of the biologically important Caesium-137 (half-life 30 years) and Strontium-90 (half-life 28 years), a test series (the bulk of which) was spread over 20 years) is little different from a war in which the detonations all occur over a much shorter time. The *Ambio* scenario war would result only in about three times as much long-lived fallout reaching the Southern Hemisphere as occurred during and after all of the test series.

Three strong qualifications need to be placed on this conclusion. One is that the Southern Hemisphere fallout could be considerably greater if large quantities of fission products are carried into the stratosphere by urban fire plumes or convection driven by solar heating as discussed in section (ii) above, or if large quantities are carried into the southern troposphere by anomalous monsoon-type

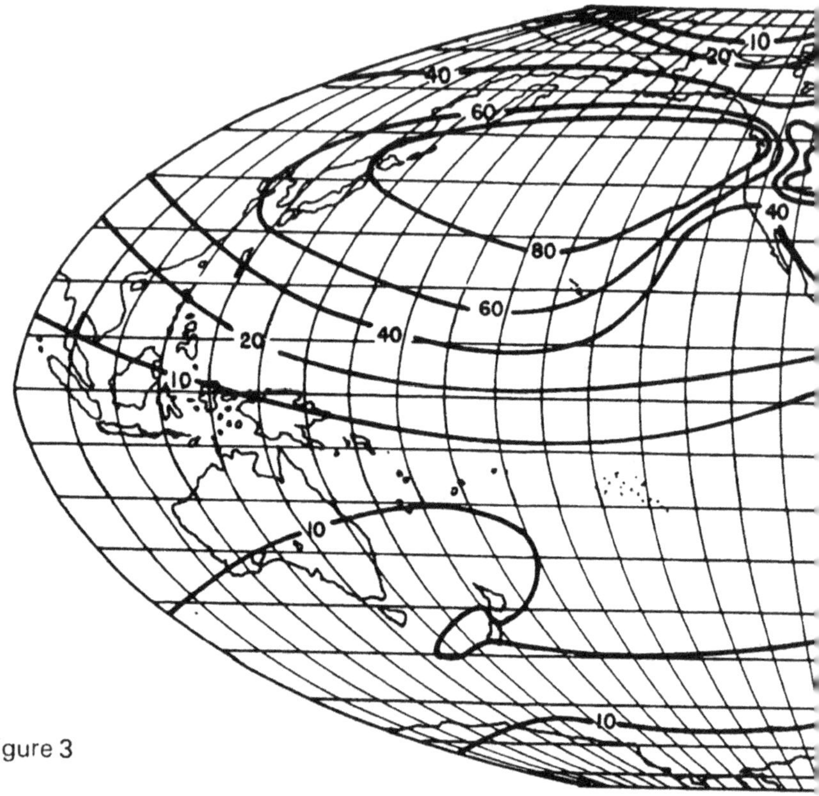

Figure 3

Map of cumulative Strontium-90 deposition from atmospheric nuclear bomb tests, in mCi/km², based on analyses of soils collected in 1965-67. After Hardy *et al.* (1968)

circulations as discussed in section (iv) above. These possibilities are not allowed for in Enting's calculations.

A second major qualification is that, while the ten-fold increase in total yield in an *Ambio* scenario war compared to the test series may only result in a three-fold increase in long-term fallout via the stratosphere, there would be a correspondingly greater increase in the amount of early and delayed fallout in the northern troposphere. This is because a large proportion of atmospheric tests were conducted as air-bursts so as to minimise the amount of early or local fallout, and with warheads larger than those now likely to be used in a real war.

The third major qualification concerns local tropospheric fallout

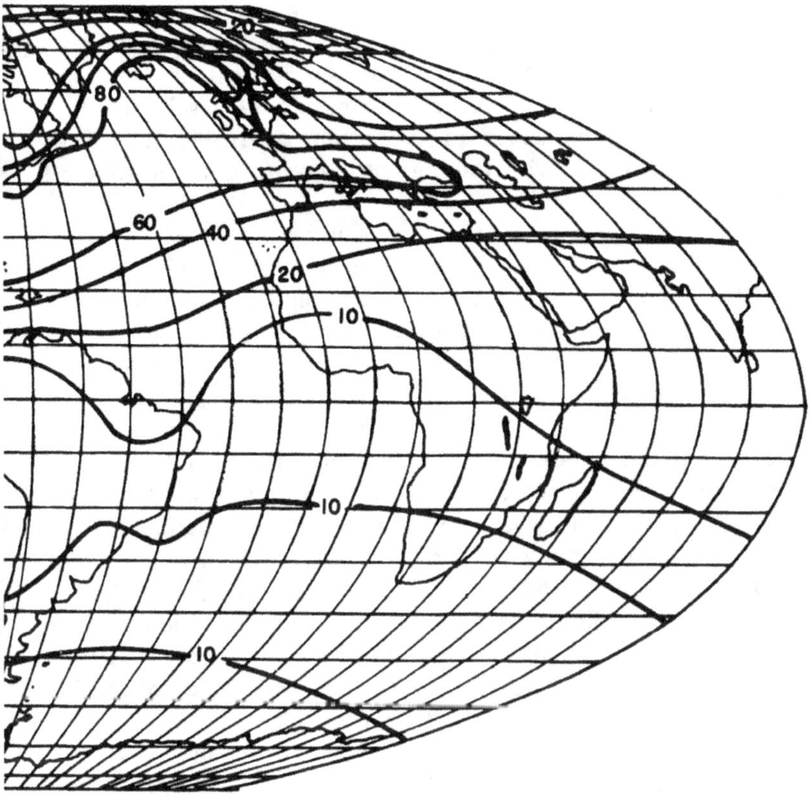

from nuclear explosions in the Southern Hemisphere. Depending on the location, size and number of such detonations these could well dominate the total nuclear fallout on Australian cities. Clearly, if this were to happen, local fallout would dwarf the contribution to Australian total fallout from detonations in the Northern Hemisphere. As far as radioactive fallout in Australia is concerned, the nature and probability of any direct nuclear attack on Australia is the critical question. This is not the case, however, for long-lived effects on climate and the ozone layer, for which transport of contaminants from the Northern Hemisphere will be of primary importance.

Bearing these caveats in mind, it is instructive to look at the way radioactivity from the atmospheric test series was distributed. Figure 3 is a map showing the cumulative concentrations of Strontium-90 arising from the test series as determined by soil measurements in 1965-67 (Hardy, Meyer, Allan and Alexander, 1968). Due to the concentration of early fallout in the latitudes where the bombs were detonated, and to the slow mixing into the Southern Hemisphere, in general Strontium-90 levels in the Southern Hemisphere reached only about one-quarter to one-third of those in the north, with the highest concentrations in middle latitudes.

Using an atmospheric transport model described by Hyson, Fraser and Pearman (1980) and the source data compiled by Enting (1982), Hyson and Enting have produced a computer simulation of the transport of radiocarbon (C-14) produced by the test series. C-14 is somewhat different to the fission products in that it is produced by high-energy neutrons in amounts roughly proportional to the total bomb yield, i.e. fission and fusion energy, whereas the fission products come only in proportion to the fission energy. Moreover, C-14 is not removed by raindrops as are many fission products and other bomb-generated pollutants. It is eventually removed mainly by dissolution into the oceans.

Figures 4(a-d) are some frames produced by Enting from Hyson's computer simulations. These show major injections of C-14 into the northern stratosphere in August and October of 1956 (Figures 4a and b respectively) and the gradual dispersion of much higher concentrations of C-14 throughout the northern and southern stratospheres in December 1962 (Figure 4c) and February 1963 (Figure 4d). Note the instrusions of C-14 rich air into the southern troposphere in the southern winter-spring period of 1956 and into the northern troposphere in the northern winter and spring of 1962-63. Stratospheric air is normally transported into the troposphere in the vicinity of the jet stream in late winter and spring.

(vii) *A Cautionary Note*

We are in fact discussing a hypothetical situation unlike, or at least different in magnitude from, any which we have observed or have 'hard data' on. We have been forced to draw upon and extrapolate from somewhat analogous situations such as urban photochemical smog, aircraft pollution effects, volcanic eruptions, and a limited understanding of climatic variability derived from data within a much narrower range of variability. Computer modelling of various components of the overall climate system can be invoked, but these are limited by uncertainties in specifying the necessary input data, by numerous simplifications in the models themselves, and by the extremities of the hypothesised situation. It is a little akin to using computer models to simulate the atmosphere of another planet. Indeed we can draw some comfort from the fact that computer models have been used with some success to simulate the behaviour of the atmospheres of Mars, Venus and the other planets (e.g., see Pollack, Leovy, Greiman and Mintz, 1981). The experience of the atmospheric bomb tests series may in some respects be our best guide but, as has been discussed, even this may not be an accurate guide to the extremities of a war situation in which the atmospheric circulation may change significantly.

The essential problem in trying to anticipate the effects on the atmosphere of a major nuclear war is in moving from a qualitative description of the factors and mechanisms involved, such as has been presented above, to a quantitative description. Besides our incomplete knowledge of many physical processes and of quantitative considerations such as average fuel densities in particular cities, what would actually happen in a future nuclear war would depend on particular circumstances such as where, why and how it were to start, shifts in alliances and degrees of support, the chance disposition of mobile forces, and the order and accuracy of particular weapon strikes.

Basic Uncertainties

(i) *In the War Scenarios*

These uncertainties include:

- the size and location of the war,
- the size, precise geographical location, and altitude of individual nuclear explosions,
- the changing nature of the weapons systems and strategies, which render yesterday's best estimates largely irrelevant for a war fought today or a decade hence,
- the time of year, which is critical to the state of the atmospheric circulation, the vertical stability of the unperturbed atmosphere, the possible spread of fire and availability of fuel, and to the seriousness of effects on crops, food and fuel supplies, and on people.

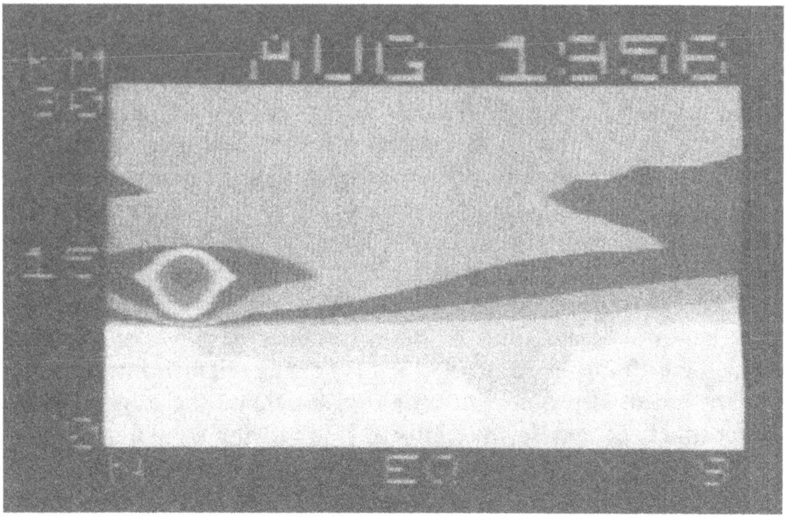

Figure 4 A

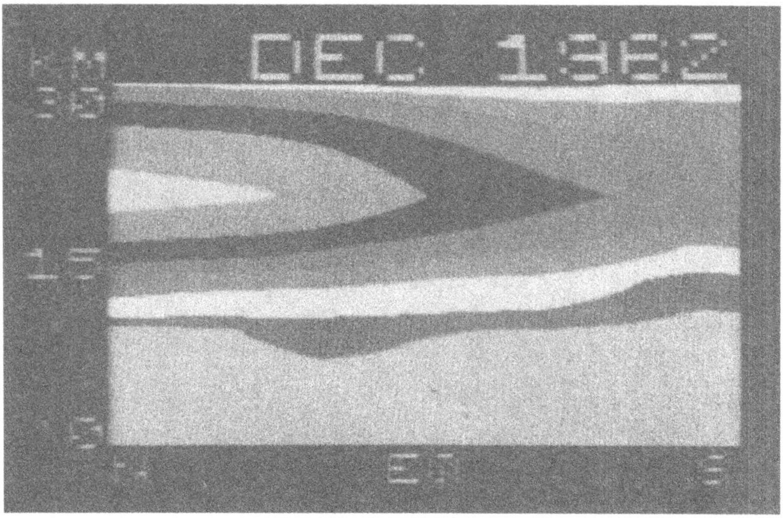

C

Selected frames produced by I. Enting (CSIRO) from a computer simulation by P. Hyson (CSIRO) of the transport of nuclear bomb test-generated radiocarbon (C-14), showing excess C-14 concentrations (over normal background levels) as a function of height (0 to 30 km) and latitude (north pole to south pole). Each successive contour change represents a doubling of C-14

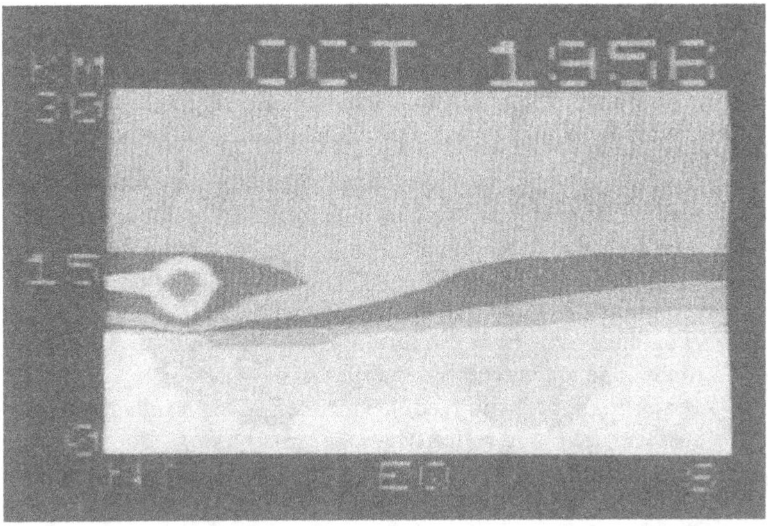

B

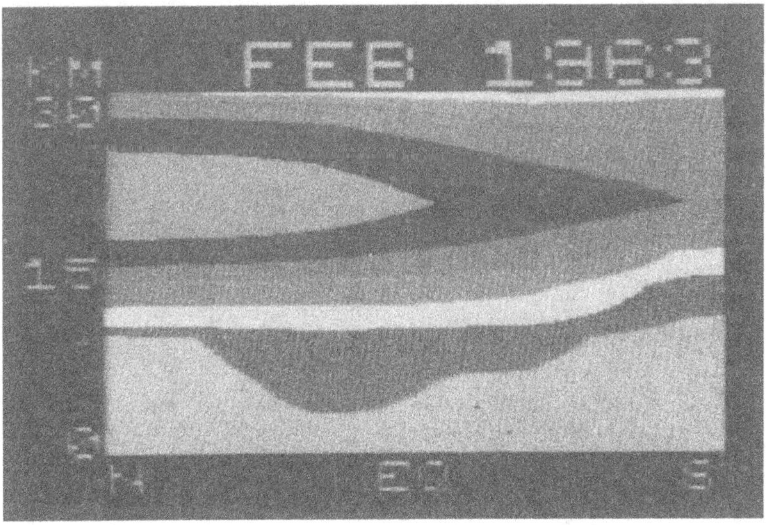

D

concentration. C-14 is produced by fast neutrons in proportion to the total yield of the nuclear explosions and is removed only by being dissolved in the oceans. It has a radioactive half-life of 5730 years and a mean residence time in the atmosphere of about 10 years. Most nuclear tests in the atmosphere took place in the tropics of the Northern Hemisphere.

(ii) *In Understanding the Atmospheric System*

Many aspects of the weather and climate system are not at present well understood. Some of these may be critical to the effects of nuclear war. Examples include:

— the microphysics of clouds and precipitation, especially as regards the effects of contaminants and additional cloud condensation or ice nuclei,
— the physics and chemistry of rainout of atmospheric contaminants, including the formation of acid rain,
— coagulation processes in dust and smoke clouds as a function of altitude and particle properties,
— climatic stability and feedback processes, especially the role of surface and cloud reflectivity and absorbtivity,
ocean-atmosphere interactions, especially involving anomalous winds and ocean currents,
— the validity of extrapolations from lesser events such as smoke plumes, volcanic eruptions, or urban pollution,
— the validity of analogies with other more distant situations such as the climates of other planets, or paleo-extinction events.

(iii) *In Understanding the Seriousness of Atmospheric Events*

For instance, given estimated effects such as high levels of photochemical smog or of UV-B radiation, how important are:

— the medical effects on humans,
— the effects on agriculture,
— the effects on oceanic food chains and organisms,
— the effects on land biota.

(iv) *Synergistic Effects*

How important are the interactions between several simultaneous stresses on the biosphere, including humans? Such stresses might include (in various combinations) nuclear radiation, UV-B radiation, reduced light levels, smog, fire, disease, polluted water and food, malnutrition, extreme weather conditions, increases in particular pests, plagues and epidemics.

Some Concluding Remarks

There can be no definitive conclusions, before the event, as to the atmospheric effects of a major nuclear war, especially in respect to effects on the Southern Hemisphere of a war conducted mainly in the north. Suffice it to say that the as yet hypothetical situation is fraught with dangerous possibilities, which may well threaten human survival in the Northern Hemisphere, and from which we in the south may not escape unscathed.

If the environmental consequences of nuclear war can be shown to be bad enough, even those people who at present find the threat and risk of nuclear war acceptable may be convinced that nuclear war is not a rational option. This implies, whether we like it or not, that the action or inaction of scientists on this issue has political implications, as do their conclusions. Inevitably, this must raise the question as to whether scientists can (or even should) be objective on such an issue.

In any practical scientific investigation the choice of questions to ask and to answer, the initial assumptions made, and many of the judgements and choices made along the way are subject to the inclinations, prejudices and biases of the individual scientists, or their employers. This is abundantly evident in the relatively non-political field of the relationship between sunspot cycles and weather or climate. The reality of this relationship has divided scientists, largely along disciplinary lines, for over a century, with high degrees of emotional involvement, biased selections of data, fallacious statistics, and erroneous logic (see Pittock, 1978, 1980a and 1983). Similar disputes arise, somewhat more understandably, in relation to most environmetal issues, be they the threat of fluorocarbons to the ozone layer (Dotto and Schiff, 1978; Pittock, 1980b), the carbon dioxide-climate problem (National Research Council, 1982b; Idso, 1982) or the safety of nuclear power (Falk, 1982).

We should be aware of similar subjective biases in assessing the seriousness of the environmental effects of nuclear war. To take some well-known extremes, one might contrast the public statements of Edward Teller (see for example his article in *Reader's Digest,* December 1982) with those of Helen Caldicott in her book *Nuclear Madness.* My point, however, is not that such extreme examples exist, but that we tend to bend our investigation and understanding of the truth, especially when it is fraught with uncertainty and has large value-laden consequences, towards what we consciously or unconsciously believe or desire.

If we are to make soundly based decisions all of us as scientists and/or citizens need to attempt to examine the technical/scientific issues objectively, to identify our assumptions, biases and motivations, and to clarify and narrow down the uncertainties as far as possible.

My personal view is that as *citizens* such of us must bring our moral judgements and indeed our emotional commitment to bear on the problem of nuclear war, which in its essence is not a scientific problem but rather is 'trans-scientific' (see Weinberg, 1972 and Pittock, 1983). In this process, however, it will not do us any good to distort the scientific facts.

Our sense of morality and our emotional commitment to life may well lead us to address different questions to those asked by the military and the politicians as they contemplate and plan for nuclear war. If military strategists are thinking about and indeed are planning the 'unthinkable', then we must think about the unthinkable also. War is too serious to leave to the military — its our future that is at stake. Nuclear war is not, and indeed must not be inevitable. However it may well require an unprecedented level of commitment on our part to see that nuclear war does not occur. Nuclear war is an experiment no nation can afford to undertake.

Acknowledgements

My thanks are due to numerous scientific colleagues, and especially to Ian Enting, Ian Galbally, Peter Manins and Mary Voice for numerous discussions and permission to use their unpublished results. The views expressed in this paper do not necessarily reflect the views of the CSIRO.

References

Ambio (1982), *'Nuclear War: The Aftermath'*, *11*, Number 2-3. Royal Swedish Academy of Sciences, Stockholm.

Ambio Advisors (1982), 'Reference scenario: How a nuclear war might be fought'. *Ambio*, *11*, pp.94-9.

Batten, E. S. (1966), 'The effects of nuclear war on the weather and climate'. RAND Corp., Santa Monica, Calif., 63pp.

Broad, W. J. (1981), 'Nuclear pulse I, II and III'. *Science, 212,* 1009, 1116 and 1248.

Caldicott, H. (1978), *Nuclear Madness: What You Can Do.* Jacaranda Press, Brisbane, Qld., 120pp.

Calvert, J. G. (1982), 'The chemistry of the polluted troposphere', in Georgii, H. W. and Jaeschke, W. (eds.), *Chemistry of the Unpolluted and Polluted Troposphere,* D. Reidel Pub. Co. pp.425-56.

Carter, M. W. and Moghissi, A. A. (1977), 'Three decades of nuclear testing'. *Health Physics, 33,* pp.55-71.

Coakley, J. A. Jr., Cess, R. D. and Yerevich, F. B. (1983), 'The effect of tropospheric aerosols on the earth's radiation budget: A parameterization for climate models'. *J. Atmos. Sci., 40,* pp.116-38.

Crutzen, P. J. (1970), 'The influence of nitrogen oxides on the atmospheric ozone content'. *Quart. J. Roy. Met. Soc., 96,* pp.320-5.

—— (1979), 'The Role of NO and NO_2 in the Chemistry of the Troposphere and Stratosphere.' *Ann. Rev. Earth Planet Sci., 7,* pp.443-72.

Crutzen, P. J. and Birks, J. W. (1982), 'The Atmosphere After a Nuclear War: Twilight at Noon', *Ambio 11,* pp.114-25.

Dotto, L. and Schiff, H. (1978), *The Ozone War.* Doubleday, Garden City. N.Y., 342 pp.

Ehrlich, P. R., Ecology. Chapter nine, in J. Leaning and L. Keyes (eds.), *The Counterfeit Ark: Crisis Relocation and Nuclear War,* Ballinger (Boston), and Harper and Row (New York), (in press).

Enting, I. G. (1982), 'Nuclear weapons data for use in carbon cycle modelling', CSIRO Div. Atmos. Physics Tech Paper No. 44 (Melbourne) 18pp.

Falk, J. (1982,), *Global Fission: The Battle over Nuclear Power,* Oxford University Press, pp.410.

Gerstl, S. A. W. and Zardecki, A. (1982), 'Reduction of photosynthetically active radiation under extreme stratospheric aerosol loads', Geol. Soc. America, Special Paper 190, pp.201-10.

—— (1983), 'The Extinction of Life Due to Extreme Stratospheric Dust Loads'. Submitted to *Nature.*

Glasstone, S. and Dolan, P. J. (1977), *The Effects of Nuclear Weapons,* 3rd edn. US Dept. of Defence and Dept. of Energy.

Grobecker, A. J., Coroniti, S. C. and Cannon, R. H. Jr. (1974), CIAP Report of Findings: The Effects of Stratospheric Pollution by Aircraft. US Dept. of Transportation, Washington, DC, Rept. DOT-TST-75-50.

Hansen, J. E., Wang, W.-C. and Lacis, A. A. (1978), 'Mount Agung eruption provides test of a global climatic perturbation', *Science, 199,* pp.1065-8.

Hardy, E. P., Meyer, M. W., Allen, J. S. and Alexander, L. T. (1968), 'Strontium-90 on the earth's surface'. *Nature, 219,* pp.584-7.

Hutton, D. R. (1982), 'Atmospheric ionisation and electromagnetic pulse (EMP) effects of nuclear weapons'. *The Austr. Physicist, 19,* pp.196-7.

Hyson, P., Fraser, P. J. and Pearman, G. I. (1980), 'A two-dimensional transport simulation model for trace atmospheric constituents', *J. Geophys. Res. 85C,* pp.4443-55.

Idso, S. B. (1982), 'Carbon Dioxide: Friend or Foe?' IBR Press, Tempe AZ. 92pp.

Johnston, H. S. (1971), 'Reduction of stratospheric ozone by nitrogen oxide catalysts from supersonic aircraft exhausts', *Science, 173,* pp.517-22.

Joint Committee (1981), *Threats to Australia's Security: Their Nature and Probability,* Joint Committee on Foreign Affairs and Defence. Aust. Govt. Pub. Service, Canberra.

Lenoble, J., Tanre, D., Deschamps, P. Y. and Herman, M. (1982), 'A simple method to compute the change in Earth-atmosphere radiative balance due to a stratospheric aerosol layer', *J. Atmos. Sci., 39,* pp.2565-76.

Manins, P. C. (1983), Cloud heights and stratospheric injections resulting from a thermonuclear war', CSIRO Division of Atmos. Res. manuscript, to be published.

National Academy of Sciences (1975), *Long-Term Worldwide Effects of Multiple Nuclear-Weapons Detonations,* US National Academy of Sciences, Washington DC.

National Research Council (1982a), *Causes and Effects of Stratospheric Ozone Reduction: An Update,* National Academy Press, Washington, DC.

— (1982b), *Carbon Dioxide and Climate: A Second Assessment,* National Academy Press, Washington, DC.

Nof, D. (1983), 'On the response of ocean currents to atmospheric cooling', *Tellus, 35A,* pp.60-72.

Office of Technology Assessment (1980), *The Effects of Nuclear War.* Croom Helm,London, p.151.

Peterson, K. R. (1970), 'An empirical model for estimating worldwide deposition from atmospheric nuclear detonations', *Health Physics, 18,* p.357.

Pittock, A. B. (1978), 'A critical look at long-term sun-weather relationships', *Rev. Geophys. Space Phys., 16,* pp.400-20.

— (1980a), 'Enigmatic variations', *Nature, 283,* pp.605-6.

— (1980b), 'Monitoring, causality, and uncertainty in a stratospheric context', *Pageoph, 118,* pp.643-61.

— (1982), 'Nuclear explosions and the atmosphere', *The Austr. Physicist, 19,* pp.189-92.

— (1983), 'Solar variability, weather and climate: an update', *Quart. J. Roy. Met. Soc., 109,* pp.23-55.

Pollack, J. B., Leovy, C. B. Greiman, P. W. and Mintz, Y. (1981), 'A Martian general circulation experiment with large topography', *J. Atmos. Sci., 38,* pp. 3-29.

Pollack, J. B., Toon, O. B., Ackerman, T. P., and McKay, C. P. (1983), 'Environmental effects of an impact-generated dust cloud: implications for the Cretaceous-Tertiary Extinctions', *Science, 219,* pp.287-9.

Schell, J. (1982), *The Fate of the Earth,* Jonathan Cape/Picador/Knopf.

Shukla, J. and Mintz, Y. (1982), 'Influence of land-surface evapotranspiration on the earth's climate, *Science, 215,* pp.1498-501.

Teller, E. (1982), 'Facts and fantasies about nuclear arms', *Readers Digest* (Austr. edn) December, pp.61-5.

Turco, R. P., Toon, O. B., Park, C., Whitten, R. C., Pollack, J. B. and Noerdlinger, P. (1981), 'Tunguska Meteor Fall of 1908: Effects on Stratospheric Ozone', *Science, 214,* pp.19-23.

Turco, R. P., Toon, O. B., Pollack, J. B. and Sagan, C., (1982), 'Global consequences of nuclear "warfare",' EOS *63,* 1018, Abstract.

Twomey, S. (1977), *Atmospheric Aerosols,* Developments in Atmospheric Science, 7, Elsevier, Amsterdam, 302pp.

Westing, A. H. (1977), *Weapons of Mass Destruction and the Environment.* Taylor and Francis, (London).

Westing, A. H. (1981), Environmental Impact of Nuclear Warfare, *Environ. Conserv., 8,* pp.269-73.

Weinberg, A. M. (1972), 'Science and trans-science'. *Minerva, 10,* pp.209-22.

9 SOME CHANGES IN THE ATMOSPHERE OVER AUSTRALIA THAT MAY OCCUR DUE TO A NUCLEAR WAR

I. E. Galbally, P. J. Crutzen and H. Rodhe

Ian Galbally

Introduction

This review discusses the environmental (particularly atmospheric) effects on Australia of a major nuclear war and attempts to identify those areas about which there is at present relatively little objective information. There have been two major studies of the atmospheric effects of nuclear war (NAS 1975, Crutzen and Birks 1982) but neither has given detailed attention to the effects in the Southern Hemisphere. It is this latter aspect which is emphasised here.

Any evaluation of the consequences of a nuclear war requires some input as to the nature of the war. More specifically how many, what sized nuclear bombs are exploded where? Four nuclear war 'scenarios' are considered here. These are:

1. The Ambio scenario (Advisors 1982)
2. The Ambio scenario without nuclear explosions in the Southern Hemisphere
3. The Ambio scenario with attack on nuclear reactors (Advisors 1982) and

4. The scenario II of Crutzen and Birks (1982) (which corresponds with the NAS (1975) scenario).

No attempt is made to justify or question these 'scenarios'.

We review possible consequences for the atmosphere over Australia in the light of each of these scenarios considering particularly the penetration of sunlight through the atmosphere, the penetration of biologically harmful ultraviolet radiation through the atmosphere, and changes in the composition of rainwater. In making these evaluations we have in general assumed that the circulation of the atmosphere, particularly the interhemispheric exchange of air, is unchanged from that occurring in the absence of a nuclear war. There are some grounds for suggesting that the atmospheric circulation may be changed following a nuclear war. In that case the atmospheric effects on Australia could be increased by substantial amounts, though much less than those changes calculated for the Northern Hemisphere in the absence of the enhanced exchange. The impacts of nuclear war on the Northern Hemisphere are, for comparison, given throughout this paper.

We stress that the 'actual' effects of an all out nuclear war are not known because such a large scale war has not occurred. Many of the effects of nuclear explosions were not measured at Hiroshima, Nagasaki and in the subsequent weapons tests. Some of these effects can be estimated using physical theories but many relevant questions have yet to be adequately addressed.

It must be stressed that the atmosphere is a highly complex, interactive, non-linear system which is incompletely understood. The full response of the atmosphere to the perturbations discussed here is unknown. Only glimpses of possible effects can be seen. With these caveats in mind we present the following evaluations.

Fires and sunlight

One important feature of the nuclear bombing of Hiroshima and Nagasaki was the resulting fires. 'The one in Hiroshima was particularly severe, lasting for half a day. It completely consumed every combustible object within two kilometres of the hypocentre' (The point below the explosion). 'Hiroshima had about 76 000 buildings before the bomb was dropped. Two thirds were destroyed by fire' (Barnaby & Rotblat, 1982). This was the result of exploding one 12 to 15 kiloton yield bomb. The nuclear war scenarios we are considering involve exploding nearly 15 000 weapons with total yields in the range 100 to 1000 kilotons each.

Here we estimate the particle production from fires associated with a nuclear war and make some estimate of the subsequent attenuation of sunlight, or darkening, due to the smoke and dust clouds from the war. The conceptual basis of these estimates is outlined in Table 1 where the various factors contributing to the analysis are listed. The calculation of particle or gas production from the fire following a single explosion is as follows.

$$\text{Production} = \text{Fuel loading at site of explosion} \times \text{Area burnt} \times \text{Fraction of fuel burnt} \times \text{Emission factor}$$

This is then summed over the total number of weapons used in the scenario. These factors are now considered in detail.

In Table 2 the distribution of nuclear explosions according to land usage is presented. This distribution, derived from the Ambio scenario, (Advisors 1982, L. Taylor private communication), uses the original allocation of nuclear weapons on cities from that scenario. We then assume that all non-urban nuclear targets (military and industrial sites) are distributed between forests and agricultural/grassland according to the fractional area of the continents covered by these vegetation types taken from Lieth & Whittaker (1975). It is relevant (for later analysis) to comment that 75 per cent of the total explosive yield in this scenario (Table 3) is exploded within China, Europe, USA and USSR, in an area less than 20 per cent of the total area of the Northern Hemisphere.

The fuel loadings (amount of material available for combustion) for these various targets are listed in Table 4. The data are drawn from Nilsson (1970), Lieth & Whittaker (1975), Culver (1976), Barnett (1979), OTA (1980), Larson & Small (1982), Milthorp (1982). We estimate, based on these studies, that cities of population $> 10^6$ people have a fuel loading of 150 kg m^{-2} in a core area of 30 km^2 and a fuel loading of 30 kg m^{-2} in the surrounding suburban area. Cities with populations $< 10^6$ people are assumed to have a fuel loading of 80 kg m^{-2} in a core area of 10 km^2 and a fuel loading of 30 kg m^{-2} in the surrounding suburban area. We assume that the suburban area is larger than the area burnt.

It is difficult to estimate the area around each explosion that will burn. We assume that 50 per cent of all combustible material within the zone experiencing a thermal radiation load from the initial nuclear generated flash of $>$ 100 J cm^{-2} will be combusted in a severe fire following the explosion. Some characteristic effects of blast and thermal radiation within this zone are:

1. unreinforced brick and wood frame houses would be destroyed
2. all trees would be uprooted and broken up
3. clothing, household combustibles, dead wood and litter will ignite spontaneously, and
4. some of these small fires will be extinguished by the blast wave while others will be scattered and spread by it (Glasstone 1962, Glasstone & Dolan 1977).

To correct for overlap of areas estimated as burnt from individual explosions when multiple weapons are exploded near the one target we arbitrarily reduce our estimate of the total area burnt by 25 per cent. Our assumptions concerning the area burnt are conservative compared with the area of the Hiroshima fire, and other estimates (Broide 1960, McNea 1960, ACDA 1979, OTA 1980). The areas combused range from 20 to 240 km^2 for surface explosions of 100 to 1000 kilotons (Glasstone 1962). The area covered by the 'fireball' is 50 to 100 times smaller.

Along with the devastation of these areas, most oil storages will also be destroyed. This amounts to around 10^{11}kg (as C) in the Southern Hemisphere and 1.4×10^{12}kg (as C) in the Northern Hemisphere (Crutzen & Birks 1982).

We assume, on the basis of other large fires (Broide 1960, McNea 1960, Taylor, Corke, King, MacArthur, Packham, and Vines 1971, McArthur 1978, Glasstone & Dolan 1977), that the time taken for these fires to burn is $\sim$ 1 hr for grassland and 3 to 24 hr for forests and cities.

The emission factors for particulate material, nitrogen oxides and sulfur gases produced in these fires come from a review of many measurements of these factors presented in the literature (Vines, Gibson, Hatch, King, MacArthur, Packham and Taylor, 1971, EPA 1973, NAS 1976, Evans, King, MacArthur, Packham and Stevens 1976, Evans, Weeks, Eccleston and Packham, 1977,

Day, MacKay, Nadeau and Thurier, 1978, Siegel 1980, Cullis & Hirschler 1980, Howes, Hill, Smith, Ward and Herget 1981, Wolff & Klimisch 1982, Onnermark, Jansson, Altvall and Halvarrson 1983, Logan 1983). The average particulate emission factor obtained from this survey, 20 g particulate per kg fuel burnt is smaller than that used by Crutzen & Birks (1982) of 75g particulate per kg fuel burnt.

There are other sources of particulate material that may occur during the initial exchange. Within the nuclear fireball all materials are vaporised, including combustible materials, metals, concrete and soil. However there is only a limited amount of O_2 available in the fireball as defined by its air volume. Therefore in some cases within the fireball the oxygen demand may exceed the oxygen supply. This occurs for a ground burst of a 100 kiloton weapon over a region with a combustible loading of 70 kg/m^2. This oxygen deficiency would be much larger for an equivalent attack on a fossil fuel storage area where the combustible fuel loading must be enormous. A full storage tank 30 m high would have a fuel loading of $> 10^4$ kg m^{-2}.

Any carbon left unoxidised when all oxygen in the fireball is consumed will form soot. Then depending on the method of cooling of the fireball, or whether this soot is exposed to oxygen before or after it has cooled below 500°C, the soot will be either oxidised to CO and CO_2 or pass into the atmosphere as soot.

There could be 5×10^{11} kg of combustible material in urban fireballs and perhaps twice this amount in fossil fuel storage. Therefore the occurrence of this 'soot producing' effect, even to some small extent, could significantly increase the particulate release to the atmosphere.

The other source of atmospheric particulate material is from soil dust that is vaporised and recondensed or merely raised during the explosion. The NAS (1975) report suggests that 10^3 to 10^4 tons (10^6 to 10^7 kg) of submicron material are produced per 1000 kiloton nuclear yield. This is consistent with Israel & Ter-Saakov's (1974) estimate of 200 tons of fused soil in the fireball per kiloton yield given that this latter estimate represents all sizes of particles. This NAS (1975) estimate of submicron particles produced by the explosion represents about 10^{-4} of the soil removed from the crater by a surface burst of a nuclear weapon.

Some sources of aerosol from fires will persist after the initial nuclear exchange. When nuclear weapons are exploded as airbursts near forests then outside the zone incinerated in the initial fire following the explosion there will be a further zone where 30 per cent of the trees are uprooted and the remainder have branches and leaves blown from them (Glasstone & Dolan 1977). This devastated forest material will dry out and burn when meteorological conditions are favourable and ignition occurs.

Also there will be other areas outside the incinerated zone affected by surface bursts. This can happen where the early fallout occurs over a forest area and the radiation dose exceeds the dose required to kill the trees. The total radiation dose levels required to kill trees are ≥ 1800 roentgens for coniferous trees and ≥ 5000 roentgens for deciduous trees (Woodwell 1982). No information is available on the dose required to kill trees in tropical forests so we assume it is ≥ 5000 roentgens. When these levels are exceeded due to early fallout, most of the cumulative dose is received in a day or two of the explosion and the tree canopy will rapidly die. No doubt these forest areas also will burn as soon as conditions are favourable for combustion. We have calculated the areas affected in this way from fallout patterns (Glasstone 1962) with weighting according to the proportion of forests on each continent that are coniferous and non-coniferous, (due to the different lethal radiation doses of coniferous and non-coniferous forests). These forests are presumed to burn when meteorological conditions are conducive and accidental or deliberate ignition takes place during the six months (covering summer, autumn and early winter in the Northern Hemisphere) following the Ambio scenario war which occurs on June 10.

Another continuing source of combustion products will come from burning oil fields and gas wells attacked during the war. Crutzen & Birks (1982) have documented this, and various further comments are appropriate. Not all wells flow naturally. In Australia out of 447 producing wells only 105 flow naturally from their own pressure. Many offshore wells have automatic or remotely operated down-hole safety valves to prevent accidental releases. However these safety valves have had (in the past) a very poor performance record with 25 per cent failure rate in operational conditions (Kash, White, Bergey, Chartock, Devine, Leonard, Salomon and Young 1974). Studies of past oil spills at sea (Jernelöv & Linden 1981) indicate that 1 per cent of the oil is burnt and 50 per cent evaporates. This evaporated material consists of hydrocarbon molecules with a carbon chain length of 5 to 15. These molecules would participate in atmospheric photochemistry.

At present there are burning and leaking oil wells in the Persian Gulf due to military activities (*The Times,* London April 15 1983, *Times,* New York, April 18 1983). The largest offshore gas well blow out in Australian exploration history occurred at Petrel No. 1 located 260 km west of Darwin and 160 km from nearest landfall. The blow out and ignition occurred in August 1968 during drilling. The initial gas flow was estimated to be several million ft^3 per day ($\sim 10^5$ m^3 day^{-1}). The blow out burned at a slowly diminishing rate for about 16 months until sealed by a relief well in January 1970 (Beddoes 1973). Thus it is highly probable that there would be continuing oil and gas fires after a nuclear war. However the likely flow rates from uncapped wells are unknown.

The estimates of initial and delayed production of aerosol from the Ambio scenario nuclear war for the Southern Hemisphere and Northern Hemisphere are presented in Table 6. Around half the aerosol emission comes from city fires and the other half is made up of approximately equal contributions from dustrise, forest fires and the burning of fuel storages. The only quantifiable source of postwar aerosol emission is that due to delayed forest burning in zones killed by radioactivity. These areas are 0.5 to 1.5 $\times$ 10^4 km^2 in the Southern Hemisphere and 2 to 6 $\times$ 10^5 km^2 in the Northern Hemisphere.

The total aerosol production during the initial exchange is $\sim$ 10 $\times$ 10^{12}g in the Southern Hemisphere and 200 $\times$ 10^{12}g in the Northern Hemisphere.

We acknowledge that these estimates are uncertain, but insufficient information is available to assess the uncertainty. If none of the forest material burnt (an unlikely situation) the particulate production would be reduced by only 15 per cent. Alternatively it appears quite feasible, in the light of the figures we have examined, that the total aerosol emission could be much larger than the 'best estimate' arrived at here.

The initial distribution of this aerosol in the atmosphere may be estimated from information about the sources. The aerosol from fires will rise in the atmosphere.

We calculate this rise using conventional plume rise theory and the heat flux from the fuel combusted in the fire. This plume rise theory has been developed for a nuclear war fire scenario, Manins (1983). Typically we find for 1 Mt of total explosion on a particular target and assumed burning times of 1 hr for grassland and 3 to 24 hr for forest and cities, the top of the plume reaches 7 km for grassland and 7 to 12 km for forests and cities. The centre line of these plumes would be at $\sim$ 0.8 of the top height and the bottom of the plume would be at 0.6 of the top height i.e. the minimum height of these plumes will be around 4 km. The soil dust (submicron) will be distributed according to the final heights of the initial nuclear 'mushroom' clouds, and for Ambio Scenario I 90 per cent of the soil dust will be between 7 and 13 km. Thus the final aerosol layer will reside mainly between 4 and 13 km.

The horizontal extent of this initial aerosol layer is determined by the initial width of the plumes at their equilibrium height, the prevailing wind speed during the plume rise and the spacing between the targets (or the degree of overlap of the plumes). We assume that the plume from a fire from a 1 Mt target is typically 15 km wide at its equilibrium altitude and initially experiences a wind speed of 25 m s^{-1} at altitude (Palmen & Newton 1969). Thus the cloud size from a grass fire might be 1×10^9 m^2 and for city and forest fires between 3×10^9 m^2 and 3×10^{10} m^2. The Ambio Scenario has around 200 targets in the Southern Hemisphere and perhaps 5500 in the Northern Hemisphere, and these typically receive around 1 Mt of nuclear explosive. We estimate that the total area of cloud initially produced (neglecting overlap) would be in the Southern Hemisphere 0.2 to 2 per cent of the hemispheric area and in the Northern Hemisphere 7 to 40 per cent of the hemispheric area. In the Southern Hemisphere the question of overlap is not important because even the most extensive cover of the clouds ($\leq$ 2 per cent) is a very small fraction of the hemispheric area. However in the Northern Hemisphere the area of potential smoke and dust aerosol cloud cover (7 to 40 per cent) is sufficiently large to obscure much of the sky and so the question of overlap reducing the cloud extent is important. According to Table 3, 75 per cent of the total nuclear explosive yield will be used in China, Europe, USA and USSR whose combined land area is 18 per cent of the Northern Hemisphere. Alternatively we note that in the Ambio Scenario I, around 91 per cent of the total nuclear weapons yield is exploded between 20°N and 60°N (H. Rodhe unpublished data). Inspection of the Ambio Scenario Target Map, No. 4, indicates that the majority of targets are dispersed over the area covered by the USA, Europe, USSR west of the Aral Sea, eastern China, North and South Korea and Japan. This area, about one half of the land area between 20°N and 60°N or 12 per cent of the hemispheric area, represents a reasonable upper limit to the initial dispersion of the smoke and dust clouds during the 24 hrs following the commencement of the war, rather than the 18 per cent or 40 per cent discussed above. We estimate the aerosol loading of these clouds, from the total aerosol masses emitted presented in Table 6, to be 0.2 to 2 g m^{-2} in the Southern Hemisphere and 6–13 g m^{-2} in the Northern Hemisphere. The higher loadings in the Northern Hemisphere result from both the greater proportion of urban and forest targets in that hemisphere and the considerable overlap of plumes in that hemisphere.

The aerosol produced during and subsequent to a nuclear war will undergo transformations in the atmosphere. Here we are primarily concerned with the attenuation of sunlight (direct plus scattered) reaching the earth's surface. The attenuation of sunlight by aerosol is dependent on the refractive index of the aerosol, which determines the proportion of scattering versus absorption, on the geometric cross sections of the particles involved and on an optical extinction coefficient dependent on refractive index, particle radius and wave length (Friedlander 1977, Twomey 1977).

The peak in the size distribution of fire smoke particles appears to be around 0.05 to 0.1 μ radius (Vines et al. 1971, NAS 1976, Day et al. 1978). The smoke particles are no doubt strongly absorbing. We assume that the mixture of dust and smoke would behave with an effective refractive index of 1.5-0.1i, a value characteristic of heavily polluted urban atmospheres. Two independent approaches have been considered in relating the optical depth of aerosol to the columnar mass loading of aerosol generated by fires (Crutzen & Birks 1982, Deirmendjian 1969, Barton & Paltridge private communication). These yield an extinction optical depth of around 8 for a mass loading of 1 g m^{-2} aerosol.

The attenuation of sunlight is calculated using the parameterised scheme for radiation scattering and absorption in aerosol layers developed for thick clouds on Venus (Sagan & Pollack, 1967). Beneath the aerosol clouds which cover $\geq$ 2 per cent of the Southern Hemisphere the intensity of sunlight at noon is estimated to be at most $\sim$ 20 per cent of that on a normal day.

These clouds in the Southern Hemisphere will probably have no large environmental impact. They will be carried by winds around the hemisphere and dispersed in a few days. The total aerosol mass predicted for injection in the Southern Hemisphere lies somewhere between the mass injected by the Krakatoa (1883) and Agung (1963) volcanoes, (Deirmendjian (1973)). The climatic impact of these volcanoes, and by analogy the dust from nuclear weapons in the Southern Hemisphere, while detectable (NAS 1975) would be insignificant compared with the more direct effects of these explosions.

In the Northern Hemisphere the situation is more complex. The total aerosol mass injected from this hypothetical nuclear war is perhaps 10 times that injected by the Krakatoa volcano and more than 100 times the natural loading of the atmosphere (Twomey 1977). We calculate that the huge black clouds formed over the target areas with columnar aerosol loadings of 6–13 gm^{-2} absorb 92 per cent and reflect 8 per cent of the incoming solar radiation, and transmit virtually no sunlight to the surface. There would be immediate effects on surface temperatures in continental areas away from oceans due to this blocking of sunlight. The darkness and cold (in inland regions) combined with the general shortage of medical facilities, food and shelter, will make the task of surviving more difficult for the remaining population.

The clouds have such large optical thickness, τ = 50 to 100, that on average (assuming they were well mixed) all the absorption of solar radiation would take place in the top 1 km. This 1 km layer would experience an initial heating rate due to solar radiation of $\sim$ 100 K day^{-1} as a 24 hr average. The equilibrium temperature for this 1 km layer with this albedo would be at least 270K. The heating of the layer could cause the rapid buoyant convection of these clouds into the stratosphere. Once in the stratosphere, the lifetime of the clouds would be greatly prolonged permitting them to become dispersed over the whole globe and persisting for months to years. If 50 per cent of the aerosol emitted in the Northern Hemisphere by this hypothetical war was dispersed over the globe as an aerosol layer, its column mass loading would be 0.2 g m^{-2}, its optical depth would be $\sim$ 1.5 and it would absorb or reflect $\sim$ 80 per cent of the incoming solar radiation. As the circulation between the hemispheres is quite rapid above 20 km the Southern Hemisphere would not escape from such a global darkening event.

There are other processes which could affect the fate and the attenuation of sunlight by this aerosol layer. If these processes are rapid and effective they may modify the effects of the aerosol just described.

Processes affecting the optical depth of such clouds are:

1. the production of new aerosol particles,
2. the coagulation of aersol particles,

3. the diffusion and dispersion of the aerosol throughout the atmosphere, and
4. the removal of this aerosol by precipitation scavenging and dry deposition.

It should be stressed that these processes are interactive and that a proper evaluation of the subsequent fate of these aerosol clouds requires complex modelling not yet undertaken. Any change in the albedo or heating rate of the atmosphere will induce some change in atmospheric dynamics, cloud formation and precipitation. Obviously reduced precipitation through the aerosol clouds would increase their lifetime, whereas increased precipitation will reduce it. Changes in one direction or the other would be likely if such aerosol clouds entered the atmosphere. We believe that even the direction of such changes is presently unknown. In the absence of the modelling necessary to quantify these processes we attempt to critically assess the time scales of processes 2, 3 and 4 in an *unperturbed atmosphere* and their likely effect on the attenuation of sunlight by the clouds.

The processes of coagulation and dispersion of aerosol are coupled because the coagulation rate is dependent on the square of the aerosol concentration. So dispersion of aerosol into clear air reduces the total coagulation rate. Furthermore, for a constant volume (or mass) of aerosol the optimum size aerosol for optical extinction is 0.25 μ radius (see Friedlander 1977, p.135). The predominant size particles in fresh smoke is 0.05 μ radius. It takes 125 particles of 0.05 μ radius to make up the volume of one 0.25 μ radius particle, so substantial particle number reductions can occur while the optical depth of the smoke may even increase! We have previously calculated that the fresh smoke clouds have mass loadings of 6 to 13 gm^{-2} distributed over 9 km depth. This corresponds with particle densities of 2 to 5 $\times$ 10^4 particles/cm^3 with a peak number density at 0.05 μ radius (Vines et al. 1971, Barton & Paltridge private communciation).

Simple coagulation theory (Friedlander 1977, Twomey 1977) indicates that at these initial concentrations $\sim$ 3 days are required for a factor of 10 decrease in particle number concentration and $\sim$ 30 days for a further factor of 10 decrease. We cannot assess the exact influence on optical depth of this particle number decrease as it requires complex coagulation calculations but considering the discussion above it is not obvious that the optical depth would greatly decrease during the first week or so of coagulation (e.g. see the aerosol distributions in Burgmeier, Blifford and Gillette 1973). Furthermore the abovementioned coagulation times would be lengthened by dispersion of this aerosol into the stratosphere or through the troposphere.

Simultaneously with this coagulation, there will be dispersion of these aerosol clouds throughout the atmosphere. The rate of dispersion in the troposphere depends on the initial size of the clouds and these differ greatly between the hemispheres. Reasonable estimates of the time for spreading of these aerosol clouds throughout the troposphere (if they do not pass into the stratosphere) are about one month for the Southern Hemisphere and two weeks for the Northern Hemisphere.

From studies of radioactive material (Lambert, Sanak and Polian 1983), of soot particles (Ogren & Charlson 1983) and of the frequency of occurrence of clouds and precipitation (Rodhe & Isaksen 1980) we estimate the average lifetime of submicron aerosol particles in the upper troposphere to be in the range 10 to 30 days and perhaps 10 times as long in the lower stratosphere (NAS 1975). This implies that the mass of such aerosol particles produced during and immediately following the war and contained in the upper troposphere would decline due to scavenging (by precipitation) to 30 per cent within two weeks to a month. The mass of aerosol particles in the lower stratosphere would decline similarly due to stratospheric-tropospheric exchange within three to twelve months.

The steady state loading of aerosol mass in the troposphere due to forest fires, and oil and gas well fires in the months after a nuclear war would be $\sim$ 0.001 g m^{-2} in the Southern Hemisphere and $\sim$ 0.03 g m^{-2} in the Northern Hemisphere. This loading is below the natural level in the Southern Hemisphere, but somewhere between the background level and that from the Krakatoa injection in the Northern Hemisphere and as such could cause marginally detectable climatic changes.

In summary it appears that fire-smoke and dust rise will form black clouds over all target areas following a nuclear war. Initially these clouds would only affect surface temperature in the Northern Hemisphere. However it is probable that some fraction of these clouds will buoyantly rise into the stratosphere and darken the sky globally for months. Alternatively if they remain in the troposphere (and they do not perturb the dynamics of the atmosphere and the frequency of precipitation) the clouds could coagulate, disperse and be scavenged during a few weeks after the war. In the latter case, the effects of the clouds on surface temperature and the weather would be confined to the Northern Hemisphere (provided there is no change in tropospheric interhemispheric exchange). The temperature and weather changes would last perhaps no longer than the clouds themselves.

It must be recognised that there is great uncertainty in many of the figures presented. Here we have attempted to take the most reasonable or median value for any particular term. In some cases the upper and lower limits are an order of magnitude different from the values chosen. The uncertainty in our final calculations is probably at least this large.

Other Atmospheric Effects of a Nuclear War

The effects of this hypothetical nuclear war on stratospheric ozone, on radioactivity and on the acidity of rainwater are also evaluated. All these evaluations are based on the assumption of an unperturbed atmospheric circulation and the absence of the aerosol clouds already discussed. The presence of these aerosol clouds, or any effect that they have on atmospheric dynamics could profoundly alter the conclusions drawn here. Several possible changes in atmospheric dynamics due to such clouds are discussed in the accompanying paper (Pittock 1983).

Ozone absorbs incoming solar radiation of wavelength shorter than about 320 nm, and thus shields the earth's surface from biologically damaging ultraviolet radiation. A change in the ultraviolet radiation reaching the earth's surface due to a change in the amount of ozone could have undesirable effects on those biological systems that are exposed to sunlight. In addition, ultraviolet as well as infra-red absorption by ozone plays an important role in determining atmospheric temperatures and climate.

The actual distribution of ozone in the atmosphere is determined by the combined effects of its production and destruction processes (including temperature — and radiation — induced variations) along with atmospheric transport processes. Hence the only way that the theory and observations of atmospheric ozone can be compared is by means of complex numerical simulation of all the processes involved.

In the stratosphere, molecular oxygen, O_2, absorbs solar radiation and dissociates into two oxygen atoms. These oxygen atoms combine with two oxygen molecules to form two ozone molecules as follows:

$$O_2 + h\nu \rightarrow O + O$$

$$O + O_2 + M \rightarrow O_3 + M \text{ (Twice)}$$

The effect of oxides of nitrogen is to catalyze ozone destruction via the reactions:

$$NO + O_3 \rightarrow NO_2 + O_2$$

$$O + NO_2 \rightarrow NO + O_2$$

$$O_3 + h\nu \rightarrow O_2 + O$$

$$\text{Net: } 2\,O_3 \rightarrow 3\,O_2$$

It is now recognised that this cycle is the principal means by which ozone is limited in the natural stratosphere (WMO 1982, NAS 1982).

Nitric oxide is produced by nuclear weapons by the heating of air in the interior of the fireball and in the shock wave (Gilmore 1975). This nitric oxide is mixed throughout the nuclear cloud. The cloud, for bombs with total yields of 1 Mt or greater, penetrates the tropopause depositing substantial amounts of NO in the stratosphere. At heights above 20 km this NO is expected to cause ozone depletion.

The clouds from bombs of total yield smaller than 1 Mt do not penetrate deep into the stratosphere and so for a given total megatonnage of weapons a shift towards smaller weapon size decreases the effect on stratospheric ozone whereas an increase in individual weapon size increases the effect on stratospheric ozone. This factor is important in understanding differences between the various assumed scenarios (Whitten, Borucki and Turco 1975, NAS, 1975, Crutzen & Birks, 1982).

It is not expected that there would be any significant direct effect of nuclear weapons exploded in the Southern Hemisphere on the ozone layer. The 173 Mt of weapons exploded in the Southern Hemisphere in the Ambio Scenario I (see Crutzen & Birks 1982) is smaller than the 300 Mt of mainly high yield bombs used in atmospheric tests by the US and USSR in 1961 and 1962. There has been considerable debate as to whether these bombs produced an ozone decrease of a few percent (Chang, Duewer and Wuebbles,

1979). Because of the large scatter in ozone measurements and our lack of understanding of all of the natural causes of ozone fluctuations, it has not been possible to unequivocally identify an ozone decrease due to these weapons tests. Thus the direct effect on the stratosphere of the 175 Mt allocated to the Southern Hemisphere in the Ambio Scenario I would probably be undetectable irrespective of the yields of the weapons used.

Crutzen & Birks (1982) consider the influence on stratospheric ozone of nitrogen oxides injected into the stratosphere from two different war scenarios. In Ambio Scenario I, the weapons are primarily low yield and no significant ozone depletion occurs. However, in the Scenario II, where high yield weapons are used there is massive injection of NO_x into the Northern Hemisphere stratosphere, perhaps twenty times the natural level, and we must consider the effect of this and its spread to the Southern Hemisphere.

Model studies of ozone depletion from this type of scenario have been carried out by Whitten et al. (1975), Chang (see NAS 1975) and more recently by Crutzen & Birks (1982).

Crutzen & Birks (1982) two-dimensional model predicts a rather uniform 65 per cent depletion of the ozone column spread from 45°N to the North Pole by the 50th day following the war. The depletions become less toward the equator and beyond, being 57, 42, 26, 12 and 1 per cent at 35°N, 25°N, 15°N, 5°N and 5°S, respectively. As time progresses, the ozone depletions become less in the Northern Hemisphere, but NO_x is transported to the Southern Hemisphere and causes significant depletion there. Two years following the war in the Northern Hemisphere the ozone column depletions vary uniformly from 15 per cent at 5°N to 56 per cent at 85°N, with a 39 per cent depletion of the ozone column at 45°N. At the same time ozone column depletions range from 12 per cent at 5°S to 18 per cent at 85°S in the Southern Hemisphere.

There are some important uncertainties in these model calculations. Along with the nitrogen oxides, large quantities of water vapour and particulates will be injected into the stratosphere. These particulates could have some minor role in contributing to the ozone destruction chemistry. This cannot at present be quantified. More importantly if the particulates are light absorbing they will contribute to the local heating of the stratosphere. Ozone depletion will, of course, lead to local cooling. Some of the ozone destroying reactions are dependent on temperature so these changes in heating rates are important. Furthermore the circulation of the stratosphere is driven by latitudinal and vertical differences in heating and cooling rates. A thick layer of light absorbing aerosol in the lower stratosphere would affect the dynamics of the stratosphere, the temperature of the stratosphere, the distribution of ozone destroying pollutants and ozone depletion in complex ways which we cannot predict. We can be confident, however, that the perturbation in the ozone column of the Northern Hemisphere would be quite large for a Scenario II nuclear war. The magnitude of the effect in the Southern Hemisphere is more uncertain.

The effect at the earth's surface in the Southern Hemisphere of a 10 per cent ozone decrease has been calculated by Paltridge and Barton (1978) and Stordal, Hov and Isaksen (1982) to be in the range of a 20–30 per cent increase in biologically damaging ultraviolet radiation (UV-B). The expected adverse effects of increased levels of UV-B include increased incidence of skin cancer in fair skinned races, decreased crop yields and a variety of stresses on terrestrial and aquatic ecosystems (NAS 1979).

One factor which could mitigate these stresses from UV-B is the presence of enhanced levels of atmospheric aerosol from the nuclear weapons and subsequent fires. Evans et al. (1977) observed a 20 fold decrease in total UV radiation (direct plus diffuse) under a smoke plume that had a optical depth for scattering of $\tau_{scatt.} = 1.6$. Obviously a minor, but persistent enhancement of atmospheric aerosol in the Southern Hemisphere could mitigate the UV-B stress on biological systems. Similarly a persistent doubling of tropospheric ozone (Crutzen & Birks 1982) would negate the effects of a 10 per cent decrease in stratospheric ozone. However these effects would have to last for as long as the ozone depletion and such a coincidence is unlikely.

Another effect of such a war is the introduction of radioactive material into the atmosphere from the fission material in the bombs and also from any nuclear facilities attacked in the war, (Advisors 1982). We examine here the deposition of delayed fallout (that which occurs after 24 hr) over Australia from this hypothetical nuclear war. The deposition in the Northern Hemisphere has been extensively examined elsewhere (Ambio 1982).

A certain fraction of the radioactive material introduced into the NH atmosphere during a nuclear war will eventually find its way across the equator and be deposited in the Southern Hemisphere Because of the time it takes to mix air from one hemisphere into another — several months — we need only concern ourselves with nuclides with a half-life comparable to or longer than this time scale. In the following estimate we concentrate on ^{90}Sr (Half-life 28 years) for which there exists fallout data from the nuclear bomb test period. ^{90}Sr is also important because of its tendency to accumulate in certain parts of the human body (bones and marrow).

As a basis for the calculations we make the following assumptions:

1. Only that fraction of the radioactivity that resides in the stratospheric portion of the stabilized bomb clouds is available for transport across the equator; the tropospheric portion of the Northern Hemisphere emission is assumed to be deposited — mainly by precipitation — before the air can reach the Southern Hemisphere.
2. The deposition of the stratospheric fraction is assumed to be distributed between the latitude bands similar to the observed distribution of ^{90}Sr in soils a few years after the bomb test period in the late fifties and early sixties (Hardy, Meyer, Allen and Alexander, 1968).
3. The tropospheric fraction (excluding the local fallout) of the Southern Hemisphere bomb emissions is assumed to be deposited uniformly between the equator and 45°S.

With these assumptions and with due regard to the radioactive decay the following deposition values result (Table 7). Ambio's Scenarios I would result in an increase in the average ^{90}Sr deposit of about a factor of five. If, in addition, all nuclear reactors were hit, each by a 1 Mt bomb, the ^{90}Sr deposition would rise by another factor of 6. The deposition values in Table 7 represent averages for the latitude band 10°S-40°S. Substantial deviations are expected to occur mainly in connection with differences in precipitation amounts. The relation between the ^{90}Sr deposit in Southern Hemisphere soil samples during 1965–1967 and the average annual precipitation has been examined. Despite a considerable scatter in the data, probably partly due to deviations from average rainfall during the particular years in question, a higher deposit is associated with high average rainfall.

We estimate that certain high rainfall areas may receive at least a factor of three higher deposits than those indicated in Table 7. For the Scenario that includes nuclear reactors certain sites — particularly in the mountain regions on the east coast of Australia — may thus receive a ^{90}Sr deposit of roughly 1 Ci/km^2.

The deposition of ^{137}Cs would be similar but 50 per cent higher than that of ^{90}Sr (because of a higher yield of ^{137}Cs in the fission process).

Considerable amounts of sulfur and nitrogen oxides will be introduced into the atmosphere as a result of the fires during a large scale nuclear war. These oxides, particularly when further oxidised to sulfuric and nitric acid, will tend to make aerosols, cloud droplets and precipitation water acidic. In the bomb clouds from surface bursts the acidity so produced will be neutralised to a certain degree by alkaline material, e.g. calcium carbonate and metal oxides originating from the surface. The sulfur and nitrogen oxides emitted by the subsequent fires in urban areas, forests and oil and gas wells are less likely to be neutralised in this way.

We have made rough estimates of the acidity of precipitation during the weeks and months following the Ambio Scenario war. The following assumptions are made:

1. Thirty per cent of the oxides are deposited by direct uptake at the surface without prior oxidation to sulfuric and nitric acid. The remaining 70 per cent is deposited as acid in precipitation.
2. No neutralisation of these acids takes place in the atmosphere (this is a worst case assumption).
3. The deposition by precipitation of the sulfur and nitrogen emitted in association with the war will take place in the same latitude belt as the emission within two weeks of the emission.
4. Natural processes alone would maintain a pH in rainwater of about 5, as they do now.
5. The intensity of precipitation is unchanged.

We distinguish between short term emissions, i.e. those taking place during the first few days after the war and longer term emissions taking place during the first six months. The emissions are calculated on the same basis as aerosol emissions using the relevant emission factors in Table 5. For long term emissions we add an additional source for burning oil and gas wells based on the assumptions made in Crutzen & Birks (1982). The resulting pH values are shown in Table 8.

The pH values in Table 8 represent average values for the latitude belts indicated. As with radioactivity, certain regions within these belts may receive rain with elevated contamination. This rain may have a pH value several tenths of a unit lower than the zonal average values in Table 8. These pH values are more acidic than those experienced in some industrialised regions of the earth (Rodhe 1981).

Conclusions

It appears that the greatest atmospheric environmental hazard accompanying a nuclear war is the attenuation of sunlight by smoke clouds generated by the fires from nuclear explosions. These calculations support and extend the previous work (Crutzen & Birks 1982) which concluded that at least in the northern hemisphere there could be 'twilight at noon'. We suggest that this darkness may extend to the Southern Hemisphere. We believe that sophisticated model calculations are required to more thoroughly examine these possibilites.

Table 1

Basis of Atmospheric Aerosol and Solar Attenuation Estimates: Ambio Scenario

1. *Fires*
Land usage around targets
Fuel loading according to land usage
Area of combustion
Fraction of fuel burnt
Emission factors
Duration of fire
Plume rise
Optical properties of smoke
Solar radiation attenuation model
1. (a) Immediate Fires
Area: Flash > 100 J cm^{-2}
Duration $>$ 3-24 hr
1. (b) Delayed Fires
Area: Forests damaged by blast
Forests killed by radioactivity
Plus oil and gas wells
Duration $\sim$ 6 months as conditions conducive for ignition and combustion occur.

2. *Dust rise from surface explosions*
10^3 to 10^4 tons aerosol $< 1\ \mu$ radius per Mt yield
Vertical distribution according to nuclear cloud

Table 2

Distribution of Nuclear Explosions according to Land Usage: Ambio Scenario (Numbers of weapons used)

	Weapon Yield (kt)				
	100	200	300	500	1000
Land Usage					
Southern Hemisphere					
Cities	14	—	42	20	28
Forests	—	40	43	16	—
Crops & Grassland	—	120	130	46	—
Northern Hemisphere					
Cities	909	—	2727	500	730
Forests	5	914	484	1686	3
Crops & Grassland	11	1856	984	3424	8

Table 3

Number of Nuclear Explosions in Selected Regions: Ambio Scenario

Region	Weapon Yield 100 kt	200 kt	300 kt	500 kt	1000 kt
China	39	45	184	92	130
Europe	334	760	1353	90	138
USA	231	280	499	2234	136
USSR	199	80	864	2852	83

This represents 75 per cent of the total explosive yield allocated in the Ambio Scenario which includes:

Southern Hemisphere 173 Mt
Northern Hemisphere 5569 Mt

Table 4

Fuel Loadings kg m^{-2}

1.	Cities		
	Large Centre	0-1km	50-580
	Inner Belt	1-5 km	50-170
	Suburbs		5- 45
2.	Forests		10- 30
3.	Grassland and Crops		1- 2

Table 5

Emission Factors for Uncontrolled Fires
(g, emission/kg fuel,)

1.	Particulate Material (submicron)	
	Wild Fires (Woods etc.)	10-30
	Oil Spills (Fuel burnt)	20-50
	Plastics and Rubber	10-50
	Flare Gas	2-12
2.	Nitrogen Oxides (as NO_2	
	All cases	2
3.	Sulfur Gases (as S)	
	All cases	3

Table 6

Aerosol Emission Estimates Following a Nuclear War (Ambio Scenario)
(Tg ≡ 10^{12}g)

Source	*S.H.*	*N.H.*
1. Immediate (∿ 1 day)		
City Fires	3	90
Forest Fires	1	30
Grassland Fires	0.1	4
Oil Storage Fires	4	50
Dust Rise	0.2-2	6-60
Total	∿ 10	∿ 200
2. Delayed (∿ 6 Mo.)		
Forest Fires	1-3	30-90
Oil and Gas Wells	?	?

Table 7

Average Deposition of ^{90}Sr in Australia. (Unit: mCi/km^2).

Observed deposition in 1967 due to previous weapons tests		10
Calculated deposition using:		
1. Ambio Scenario		
Transport from N.H.	40 }	
Tropospheric fallout from S.H. bombs	10 }	.50
2. Ambio's Scenario + nuclear reactors		300

Table 8

Average Acidity (pH) of Precipitation following a Nuclear War (Ambio Scenario)

	20°S-40°S	20°N-60°N
First two weeks	4.3	3.3
First six months	4.9	4.2

Acknowledgements

We would like to thank the many colleagues who assisted with comments and advice for this work including G. P. Ayers, L. F. Evans, J. A. Heintzenberg, P. C. Manins, M. F. R. Mulcahy, J. A. Ogren, A. B. Pittock, L. Taylor, M. J. G. Wilson and many others.

References

ACDA (1979), *The Effects of Nuclear War.* US Arms Control and Disarmament Agency, US Government Printing Office, Washington DC.

AMBIO Advisors (1982), Reference Scenario: How A Nuclear War Might Be Fought, *AMBIO, 11,* pp.94-9.

Barnaby, F. and Rotblat, J. (1982), 'The Effects of Nuclear Weapons'. *AMBIO, 11,* pp.84-93.

Barnett, C. R. (1979), 'Fire Engineering Formulae for Building Designers'. *Trans. N.Z. Inst Engnrs, 6,* p.2.

Beddoes, L. R. (1973), *Oil and Gasfields of Australia, Papua New Guinea and New Zealand.* Tracer Petroleum and Mining Publ., Australia.

Broide, A. (1960), 'Mass Fires following Nuclear Attack'. *Bulletin of the Atomic Scientist, 16,* pp.409-13.

Burgmeier, J. W., Blifford, I. H. Jr. and Gillette, D. A. (1973), 'A Reinforced Coagulation-Sedimentation Aerosol Model', *Water, Air and Soil Pollution, 2,* pp.97-104.

Chang, J. S., Duewer, W. H. and Wuebbles, D. J. (1979), 'The Atmospheric Nuclear Tests of the 1950's and 1960's: A Possible Test of Ozone Depletion Theories', *J. Geophys. Res., 84,* p.1755-65.

Crutzen, P. J. and Birks, J. W. (1982), 'The Atmosphere After a Nuclear War: Twilight at Noon', *AMBIO, 11,* pp.114-25.

Cullis, C. F. and Hirschler, M. M. (1980), 'Atmospheric Sulphur: Natural and Man-Made Sources', *Atmos. Environ., 14,* pp.1263-78.

Culver, C. G. (1976), 'Survey Results for Fire Loads and Live Loads in Office Buildings', *NBS Building Science Series,* No. 85, 132pp.

Day, T., MacKay, D., Nadeau, S. and Thurier, R. (1978), 'Emissions from In Situ Burning of Crude Oil in the Arctic', *Water, Air and Soil Pollution, 11*, pp.139-52.

Deirmendjian, D. (1969), *Electromagnetic Scattering on Spherical Polydispersions*. American Elsevier, New York, 290pp.

Deirmendjian, D. (1973), 'On Volcanic and other Particulate Turbidity Anomalies'. *Advances in Geophysics, 16*, pp.267-96.

EPA (1973), *Compilation of Air Pollution Emission Factors, Second Edition*. US Environment Protection Agency, Office of Air Quality Planning and Standards, North Carolina.

Evans, L. F., King, N. K., MacArthur, D. A., Packham, D. R. and Stephens, E. T. (1976), 'Further Studies of the Nature of Bushfire Smoke'. *CSIRO Division of Applied Organic Chemistry Technical Paper* No. 2, 12pp.

Evans, L. F., Weeks, I. A., Eccleston, A. J. and Packham, D. R. (1977), 'Photochemical Ozone in Smoke from Prescribed Burning of Forests, *Environmental Science and Technology, 11*, pp.896-900.

Friedlander, S. K. (1977), *Smoke, Dust and Haze: Fundamentals of Aerosol Behaviour*, John Wiley & Sons, New York, 317pp.

Gilmore, F. R. (1975), 'The Production of Nitrogen Oxides by Low-Altitude Nuclear Explosions'. *J. Geophys. Res., 80*, p.4553.

Glasstone, S. (1957), *The Effect of Nuclear Weapons*, US Government Printing Office, 579pp.

Glasstone, S. (1962), *The Effect of Nuclear Weapons*, US Government Printing Office, 730pp.

Glasstone, S. and Dolan, P. J. (1977), *The Effects of Nuclear Weapons*, US Government Printing Office.

Hardy, E. P., Meyer, M. W., Allen, J. S. and Alexander, L. T. (1968), 'Strontium-90 on the Earth's Surface', *Nature, 219*, pp.584-87.

Howes, J. E. Jr., Hill, T. E., Smith, R. N., Ward, G. F. and Herget, W. F. (1981), *Final Report on Development of Flare Emission Measurement Methodology*, Contract 68-02-2682, Battelle, Columbus Laboratories, Columbus, Ohio, 69pp.

Israel, Y. A. and Ter-Saakov, A. A. (1974), in K. P. Makhonko and S. G. Malakhov (eds.), *Formation of Radioactive Particles during Nuclear Bursts in the Troposphere in Nuclear Meteorology*, Israel Program for Scientific Translations, Jerusalem.

Jernelöv, A. and Linden, O. (1981), 'ixtoc 1 A Case Study of the World's largest oil-spill', *AMBIO, 10*, pp.299-306.

Kash, D. E., White, I. L., Bergey, K. H., Chartock, M. A., Devine, M. D., Leonard, R. L., Salomon, S. N. and Young, H. W. (1974), *A Technology Assessment of Outer Continental Shelf Oil and Gas Operations in Energy Under the Oceans*, Bailey Bros. and Swinfern Ltd., Great Britain.

Lambert, G., Sanak, J. and Polian, G. (1983), 'Mean Residence Time of the Submicrometer Aerosols in the Global Troposphere', *Proceedings of 4th Int. Conf. on Precipitation Scavenging, Dry Deposition and Resuspension*, Santa Monica, USA, Dec 1982 in press.

Larson, D. A. and Small, R. D. (1982), *Analysis of the Large Urban Fire Environment: Part II. Parametric Analysis and Model City Simulations*, Pacific-Sierra Research Corp., Los Angeles, 56pp.

Lieth, H. and Whittaker, R. H. (1975), 'Primary Productivity of the Biosphere', *Ecological Studies 14*, Springer-Verlag, Berlin.

Logan, J. (1983), 'Nitrogen Oxides in the Troposphere: Global and Regional Budgets', *Advances in Chemistry*, Am. Chem. Soc. in press.

Manins, P. C. (1983), 'Cloud Height and Stratospheric Injections Resulting from a Thermonuclear War', Unpublished Manuscript.

McArthur, A. G. (1978), *Bushfires in Australia.* Australian Government Publishing Service, Canberra, p.359.
McNea, F. (1960), 'Fire Effects of Big Nuclear Bombs', *QUART. of the NFPA,* October 1960, pp.103-13.
Milthorp, F. L. (1982), 'Interaction of Biogeochemical Cycles in Nutrient-Limited Environments: Wheat-Pasture and Forest Systems', in I. E. Galbally and J. R. Freney (eds.), *The Cycling of Carbon, Nitrogen, Sulfur and Phosphorus in Terrestrial and Aquatic Ecosystems,* Aust. Acad. Sci., pp.35-45.
NAS (1975), *Long-Term Worldwide Effects of Multiple Nuclear-Weapons Detonations.* National Academy of Sciences, US Academy Press, Washington DC.
NAS (1976), *Air Quality and Smoke from Urban Forest Fires,* National Academy of Sciences, Washington DC, Proc. Int. Symp., Fort Collins, Colorado, Oct 24-26 1973, 375pp.
NAS (1979), *Protection against Depletion of Stratospheric Ozone by Chlorofluorocarbons,* National Academy of Sciences. National Academy Press, Washington, DC.
NAS (1982), *Causes and Effects of Stratospheric Ozones,* National Research Council, National Academy of Science, National Academy Press, Washington, DC.
Nilsson, L. (1970), *Brandbelastning i bostadslagenheter,* Statens Institut for Byggnadsforskning, Stockholm, Rapport R34, p.64.
Ogren, J. A. and Charlson, R. J. (1983), *Elemental Carbon in the Atmosphere: Cycle and Lifetime,* Tellus in press.
Onnermark, B., Jansson, R., Altvall L-E. and Halvarsson, K. (1983),'Spridning av Brandgaser och Rok mellan tva Bostadsrum', *FOA Rapport* C-20490-Dt Forsvarets Forskningsanstalt, Huvudavdelning 2, Stockholm, 63pp.
OTA (1980), *The Effects of Nuclear War,* Croom Helm, London, 151pp.
Palmen, E. and Newton, C. W. (1969), 'Atmospheric Circulation Systems: Their Structure and Physical Interpretation', *Vol 13 Int. Geophysics Series,* Academic Press, New York.
Paltridge, G. W. and Barton, I. J. (1978), 'Erythemal Ultraviolet Radiation Distribution over Australia — the Calculations, Detailed Results and Input Data', *CSIRO Division of Atmospheric Physics Technical Paper* No. 33, 48pp.
Pittock, A. B. (1983), 'The Atmospheric Effects of Nuclear War' in this volume.
Rodhe, H. (1981), 'Current Problems Related to the Atmospheric Part of the Sulphur Cycle, Some Perspectives of the Major Biogeochemical Cycles', *Scope 17,* G. E. Likens (ed.) Wiley and Sons, pp.51-60.
Rodhe, H. and Isaksen, I. (1980), 'Global Distribution of Sulfur Compounds in the Troposphere Estimated in a Height/Latitude Transport Model', *J. Geophys. Res. 85,* pp.7401-9.
Rotblat, J. (1981), *Nuclear Radiation in Warfare.* Stockholm International Peace Research Institute, Taylor & Francis, London, 149pp.
Sagan, C. and Pollack, J. B. (1967), 'Anisotrophic Nonconservative Scattering and the Clouds of Venus', *J. Geophys. Res., 72,* pp.469-77.
Siegel, K. D. (1980), 'Degree of Conversion of Flare Gas in Refinery High Flares', Ph.D. dissertation, Chemical Engineering Dept., Karlsruhe, Germany.
Stordal, F., Hov, O. and Isaksen, I. S. A. (1982), 'The Effect of Perturbation of the Total Ozone Column due to CFC on the Spectral Distribution of U.V. Fluxes and the Damaging U.V. Does at the Ocean Surface', in J. Calkins (ed), *The role of Solar Ultraviolet Radiation in Marine Ecosystems,* pp.93-107.
Taylor, R. J., Corke, D. G., King, N. K., MacArthur, D. A., Packham, D. R. and Vines, R. G. (1971), 'Some Meteorological Aspects of Three Intense Forest Fires', *CSIRO Division of Meteorological Physics Technical Paper* No. 21, 20pp.

Twomey, S. (1977), 'Atmospheric Aerosols', *Development in Atmospheric Sciences, 7,* Elsevier Scientific Publishing Company, 302pp.

Vines R. G., Gibson, L., Hatch, A. B., King, N. K., MacArhur, D. A., Packham, D. R., Taylor, R. J. (1971), 'On the Nature, Properties and Behaviour of Bush-fire Smoke', *CSIRO Division of Applied Chemistry Technical Paper* No. 1, 32pp.

Whitten, R. C., Borucki, W. J. and Turco, R. P. (1975), 'Possible Ozone Depletions following Nuclear Explosions', *Nature, 257,* pp.38-9.

WMO (1982), 'The Stratosphere 1981, Theory and Measurements', WMO Global Ozone Research and Monitoring Project. Report No. 11. World Meteorological Organisation, Geneva, Switzerland, 505pp.

Wolff, G. T. and Klimisch, R. L. (eds.) (1982), *Particulate Carbon Atmospheric Life Cycle,* Plenum Press, New York, 411pp.

Woodwell, G. M. (1982), 'The Biotic Effects of Ionizing Radiation', *AMBIO, 11,* pp.143-8.

PART THREE: THE PREVENTION OF NUCLEAR WAR

10 THE ROLE OF THE SCIENTIST
Bernard Feld

Bernard Feld

There has hardly been a time in recorded history, when scientists have not played a major role in the development of the most effective and lethal instruments of warfare (e.g., Archimedes, Da Vinci and, in more modern times, mustard gas and the submarine in World War I, and radar and, of course, the atomic bomb in World War II). And yet the awful consequences of the massive strategic fire-bombings of such targets as Dresden and Tokyo in World War II notwithstanding — it was not until the awesome, essentially instantaneous destruction of Hiroshima and Nagasaki, which brought World War II to its abrupt conclusion, that civilised people of all nations realised that war had become a dangerously anachronistic means of attempting to settle differences between major, industralised nations.

Actually and unfortunately, this realisation, while it was undoubtedly very widespread in the immediate aftermath of World War II (and, indeed, was a major impetus in the formation of the United Nations and in the early efforts to invest it with effective supranational powers) was not sufficiently intense to permit the citizens of the major powers to force their leaders to take the necessary steps to turn the United Nations into an organisation with sufficient power and sovereignty to prevent national differences from turning into armed conflicts. And so the world has slowly drifted back into a situation of international anarchy which, today, is in many respects disturbingly reminiscent of that which prevailed in the earlier parts of this century and which led inexorably to the major blood-baths of 1914-18 and 1939-45.

There are those who take a much more optimistic point of view, who point out that it is precisely as a consequence of the immense lethality of modern weapons that the major powers have managed — in spite of serious and sometimes apparently irreconcilable differences — to avoid the kind of direct confrontations that might have led to an exchange of physical violence, i.e. to World War III. And, indeed, it is probably true that the United States, the Soviet Union and the other nuclear powers are more cautious in their direct actions *viz-a-viz* each other, as a consequence of their realisation of the dangers of even a conventional but direct exchange between them escalating into a nuclear conflict. But there is at least another side to that apparently optimistic approach: one may also conclude that essentially what has happened is that the conflicts among developed nations have been shifted into the third world, where there has scarcely been a year since 1945 in which bitter clashes — many of them supported, if not 'egged-on', by one or both of the so-called superpowers — have not claimed thousands of lives and led to misery and suffering for the populations involved.

Furthermore, as the major powers continue their unabated competition in the development and deployment of new weapons and weapons systems, based on the latest scientific and technological advances, the state of their equilibrium becomes more precarious, tending towards instability. This particular tendency is a direct consequence of the universal proclivity of military planners towards the belief that any apparent advantage must be pressed to its logical conclusion, lest the other side take the failure to do so as a sign of weakness, and attempt to exploit it to its own advantage. ('The best defence is a good offence' is still accepted as a matter of gospel among military planners. This is but one of a number of bits of 'folk-wisdom' that have lost their meaning, or even had their meaning reversed, in the nuclear age. The other outstanding example is 'if you want peace, prepare for war'. And yet, each new generation continues to be raised on a diet of such calumnious nonsense.)

Consequently, what appeared (at least for a number of decades after the Soviet demonstration in 1949 of its nuclear weapons capability, and in the early '50's of its mastery of the 'art' of thermonuclear weaponry) as a nuclear 'stand-off' between the west and the east, now seems to be rapidly degenerating into a competition on both sides for the achievement of some kind or kinds of 'superiority' that might be translated into actual advantage in their striving for world economic and military hegemony.

To these dangers must be added those resulting from the proliferation of nuclear weapons capabilities, and of the weapons themselves, to many other nations. In the somewhat longer run (e.g. on, say, a 50-year time scale) the most serious dangers to the survival of the human species arise, in my opinion, from the inevitable and widespread availability of nuclear weapons to a large fraction of the world's approximately 135 sovereign nations, as well as to a variety of sub-national groups and factions which will expend the necessary efforts to develop the technical capabilities and to acquire the necessary facilities for nuclear weapons acquisition.

The problem is — and it is necessary that we all face it squarely and honestly — that nature has been very unkind to us in respect to nuclear fission: given the requisite amount of fissile material (some ten kilograms of reasonably pure plutonium or one of the lighter isotopes of uranium, e.g. U-233 or U-235) it requires only a very low level of technical competence to fashion a crude nuclear bomb of devastating power. The detonation of 10 kilograms of fissile material with an efficiency of only one tenth of one percent (what any weapons designer worth his salt would call a 'dud') would release the energy equivalent of some 200 tonnes of chemical high explosive (TNT), enough effectively to destroy the centre of a major city. Most nations have scientists and engineers with the requisite training and capability for producing such a crude weapon. And it will not be very long — once terrorist groups have understood the 'advantages' of training their brightest young recruits in nuclear science as compared to, say the law — before many sub-national, irredentist groups (e.g. the Japanese Red Army, the Armenian Liberation Army, the PLO) will have acquired this kind of capability. The key, then, to the prevention of the acquisition of one or more crude nuclear bombs by such groups is in the prevention of their acquisition of the requisite 5-10 kilograms of plutonium or highly enriched uranium.

Now, both plutonium and uranium-233 are artificially produced materials not normally found in nature. Their production, in any relevant quantities, requires the operation of nuclear reactors at high power levels, followed by a complex and dangerous process of separation from a large quantity of highly radioactive spent reactor fuel. The other possible bomb material, uranium-235, is present in normal uranium in the amount of 1 part in 140. Its separation from the heavier, uranium-238 isotope (fortunately not suitable for bombs) requires a physical, rather than chemical, procedure of generally great complexity, demanding large energy consumption. Here again, unfortunately, certain possibilities — now only in the early stages of development (ultracentrifuge techniques and high-intensity laser beams of variable wave-length) — threaten to overcome this barrier as well. So far, such facilities are available only in a relatively small number of nuclear centres in the more developed regions of the world. However, this situation is very rapidly changing — if it has not already changed in an irreversible fashion — by the acquisition in many countries of high-power nuclear reactors for power production, and of the reprocessing facilities that permit the removal of the accumulated plutonium from the spent fuel of the reactor.

Generally speaking, in the West, it is almost universally believed that this form of proliferation of nuclear power facilities is inevitable, since the large majority of developing nations are unwilling to accept a situation of continuing dependency on the developed few for their growing power needs. Since the developed nations, and their industrial enterprises, look upon the growing market for nuclear power technology in the third world as a legitimate area for economic competition, the conventional emphasis on the profit motif serves as a guarantee of the continuing proliferation of nuclear technology and plutonium-production capability into the regions of the world that have, until now, been free of nuclear weapons.

However, we may well ask whether this means that we are destined to move inexorably into a world in which the nuclear Damoclean Sword will hang perpetually over our heads, and in which salvation will depend on all nuclear-capable nations developing the same level of concern (I hesitate to use the word 'responsibility') that has in the last few decades persuaded the two nuclear giants to resist temptations to unleash their nuclear war-fighting prowess? We would all like to believe that, in fact, we can still avert such a dangerous and frightening destiny.

The main hope for the avoidance of a nuclear catastrophe arises from the fact that there is, in principle, a possible route that could — if we had the intelligence to adopt it — permit the widespread utilisation of nuclear fission power without the necessary concomitant of nuclear weapons proliferation. In outline (though obviously not in actual practice) it is quite simple: to separate completely the energy production process from its fuel reprocessing and fabrication aspects. To demonstrate what is meant by the above, consider a power-reactor system which is standardised in the sense that all the reactors use the same form of fuel element (be it natural or enriched uranium or even plutonium). Let there be a single centre which is capable of reprocessing spent fuel elements and of fabricating replacement fuel for all the reactors. That centre, whether nationally or internationally operated, would be treated, in fortress-like fashion, as a repository of what is essentially the world's most precious material. (For an American, what comes to mind is Fort Knox. Others may think in terms of storage of diamonds or platinum, but plutonium is even more precious than these, and infinitely more dangerous). Fabricated fuel elements would be transported under guard, as needed, to the various reactors; the spent fuel elements, encased in lead caskets, always under guard, would be returned to the repository. Each reactor would have a recognised claim to refabricated fuel elements utilising the material it has sent to the repository. Since essentially all reactors produce less fissile material than they consume, the system would still require a net input from the outside of natural or enriched uranium, in its equilibrium operating situation. The special case of the breeder reactor, which produces more fissile nuclei than it consumes, can be fitted into such a system if and when breeder technology should become practicable.

That such a system is technically and operationally feasible, is demonstrated by the fact that it, or one very much like it, has been adopted by the Soviet Union in its extensive and active nuclear reactor export program involving the nations of Eastern Europe. While there are a growing number of operating reactors in all the Eastern European countries, the only nuclear reprocessing and fuel fabrication facilities among the Warsaw Pact nations are located in the Soviet Union. The system not only works but, in fact, permits in Eastern Europe a higher degree of nuclear energy utilisation than in the West. Obviously, the problems, associated with the adoption of a greatly more weapons-proliferation-resistant system of nuclear energy utilisation in the non-Communist world, are much more of a political than of a technical nature.

This last conclusion might be thought to undermine the thesis of this paper — that scientists and engineers have a major role to play (as scientists and engineers) in the avoiding of nuclear weapons proliferation and, especially, of the use of such weapons in situations of conflict. On the contrary, the case for the assumption by scientists and engineers of such a role is a most compelling one.

Leaving aside the common responsibilities of all citizens (from which scientists and engineers are certainly not exempt), to ensure that their governments act in the common interest, there are a number of areas in which scientists and engineers may play a special and unique role.

A major aspect of this special role of scientists and engineers is of historical origin. Not only did the scientists take the initiative (in 1938-39, immediately after the discovery of nuclear fission) in calling to the attention of their governments the unique possibilities for the utilisation of nuclear fission in the production of a super-weapon, but they were primarily responsible for the organisation and implementation of 'crash' programs, that succeeded in producing such weapons before the conclusion of the war in the Pacific (but, fortunately for the Germans, not before their capitulation on May 8, 1945). Left in the hands of the military 'professionals', the time for the production of the first nuclear explosion would undoubtedly have been longer by at least a number of years.

Thus, in a real sense, the community of nuclear scientists was deeply involved and implicated in 'the original sin'. Acutely aware of this fact, the so-called atomic scientist community lost little time in their attempts at 'atonement'. It was they who campaigned most vigorously, immediately after World War II, for the establishment of a system of civilian control, on both the national and international levels, of future nuclear energy developments; and it was they who insisted, from the start, that the achievement of a successful plan for international control requires meaningful exchanges of information and views between scientists from the East and the West, as well as between the political leaders.

In this regard, an important post-World War II development was the Pugwash Conference on Science and World Affairs. Growing out of the Russell-Einstein Manifesto of 1955, these periodic meetings of scientists from East, West, North and South have played an important role in the preservation of nuclear sanity throughout the world. Although one cannot establish a one-to-one correspondence of cause and effect, these meetings — together with the occasional informal discussions among professionals on special issues, that they have inspired — can certainly claim a non-negligible role in the achievement of the few agreements on nuclear arms control (as well as chemical and biological weapons control) that have been accomplished to date. These include: the Partial Test-Ban Treaty of 1963, the Nuclear Non-Proliferation Treaty of 1968, the ABM-Ban and SALT Agreements of 1972 and 1978, the (now tabled) Chemical Weapons Ban, to list the most significant ones.

As noted scientists and engineers cannot claim exclusive (and possiblly not even primary) responsibility for the accomplishment of these international arrangements. But neither can their role be disregarded.

In making their contributions in such areas, scientists and engineers have been most effective when they have concentrated their efforts along the following broad lines:

1. Self-education on the issues — especially the technical ones, but the political problems as well — underlying the particular negotiation in question. Although, in the final analysis, the achievement of any given agreement requires the resolution of knotty political problems, a thorough understanding of the technical basis of these problems is certainly a necessary consideration of the achievement of a positive solution.
2. Education of their fellow scientists and engineers as to the problems involved in resolving the outstanding issues.
3. Education of their governments, through direct contact with governmental officials having primary responsibility in the areas involved, as well as general education of members of involved governmental bodies (e.g. the legislatures that need to enact or approve the required measures).
4. General education of the public in their respective nations on the vital importance of constructive nuclear arms control and to the relevance and value of particular actions under consideration by their governments (especially with regard to

issues on which special-interest groups are at work attempting to undermine the agreements at issue).

5. And, finally, the keeping alive — in spite of growing world-wide tendencies toward increasingly nationalistic approaches to international problems — of the concept, vital for the survival of humankind through the coming decades, of 'one world or none'.

These are formidable and demanding tasks, not to be undertaken lightly. But, as scientists, if the foregoing analysis of our position is found to be compelling, we have no choice but to accept the responsibilities into which our special endowment, as well as the historic development of our profession (from Socrates to Galileo to Einstein) have propelled us.

11 THE ROLE OF THE MEDICAL PROFESSION

Oleg Gavrilov

Oleg Gavrilov

We physicians, as no one else, know the price of human life. Not infrequently, dozens of doctors and nurses wage a struggle for one life, fulfilling their humane duty and remembering the words of the Hippocratic Oath: 'Whatsoever house I enter, there will I go for the benefit of the sick'. And can we, therefore, entering our patients' houses, conceal from them the truth about the danger to their life and health, arising from the very existence of thermonuclear weapons in the world?

We doctors know and remember only too well what war is. We remember, because World War I took a toll of ten million human lives, whereas World War II carried away 50 million people. The calculations made at the Soviet Committee 'Physicians for the Prevention of a Nuclear War' showed that if a nuclear war broke out in Europe, even with the use of one-tenth of the power of the nuclear weapons, 314 million Europeans would become victims of nuclear strikes.

I ask myself again and again: can we physicians, whose duty is to protect life on the earth, passively watch the activities of those who are trying to suppress the instinct of self-preservation in man and conceal the truth about the real dangers of a thermonuclear war? There is only one answer. No, we cannot.

Usually all the descriptions of the aftermath of an atom bomb explosion contain the possible destruction, approximate number of killed and affected by a blast wave, luminous and radioactive radiation. The problems of medical aid in these most difficult conditions are less discussed. It is supposed without a thought in earlier plans that either hospital, medical staff and medical depots will survive the explosion or else medical aid will be rendered from undamaged areas. This is utterly absurd and far from the truth.

By present estimation, 80 per cent of the hospital beds and medical depots, plasma and plasma substitutes, and dressing material will be destroyed in the attacked city. Over 80 per cent of the doctors and medical staff will perish as most hospitals are often situated near the city's centre, the aim of the bombardment. The surviving medicos will be unable to perform their duties, and consequently, will be unable to render aid due to psychic dysfunction, which will inevitably appear at such a catastrophe.

Treatment of the wounded in hospitals will be a difficult problem. If we suppose that 30 per cent from 180 000 burned will die or will have no need for aid because of light local burns, the treatment of the remaining 120 000 will require much equipment and personnel. Britain's experience in burn treatment during World War II showed that the treatment of 34 000 burnt required 170 000 medical workers and 8000 tons of bandage material, solutions, drugs, linen, etc. The sad experience of Hiroshima witnessed that 20 000 litres of transfusion liquid were needed for the treatment of every 100 sick. For one burn victim to recover from shock 15 to 20 litres of solutions, plasma and blood, will be required in the first two days alone.

The shelf life of plasma, blood, antibiotics, and many other medicines is very short. It is impossible to create sufficient national reserves of everything needed for treating burns and wounds. Relying on new donors and speedy production of the required materials is unrealistic under the conditions of an atomic catastrophe.

The arms race and the possibility of nuclear war are the most important factors in the environment unfavourably affecting man's psyche. (Other factors include family and working conflicts, uncertainty in the future.) Unlike the widely explored problem of acute stress, the chronic effects of these human factors have still not been sufficiently investigated. But these chronic factors in the wide range (from personal to global) are vital for man's psychic well-being.

Researchers have shown that feelings of some individuals about a possible nuclear conflict very often lead to prolonged meditation and emotional reactions. Many people have dreams about the starting of nuclear conflict and the horrors of atomic bombardments.

At the same time, observations of the behaviour of large groups, who understand the catastrophic character of nuclear war, have shown their rather weak level of motivation to active struggle for preventing atomic catastrophe.

A number of hypotheses have been suggested to explain the discrepancy between the real threat of nuclear cataclysm for mankind, and the passive attitude to this threat of the various strata of the population. Attempts have been made to apply some notions from the field of individual psychology to explain the behaviour of large groups. Under investigation has been in particular, what is known by the behavioural term, 'exclusion', as well as the phenomena of 'adaptation', 'abstraction' and others. (Usually the phenomenon 'exclusion' is taken to describe the desire of people to exclude from their consciousness, unpleasant thoughts or feelings about possible malignant diseases, inevitable death, and so on.)

Public protest against the use of atomic weapons, which developed in many countries after the bombardments of Hiroshima and Nagasaki, was a direct reaction to the horrors. Nevertheless, in subsequent decades, there has been a sizeable recession in public reaction to the nuclear threat. In spite of step-by-step accumulation of nuclear weapons, there has not been a corresponding increase in public concern about the growing threat of nuclear war.

Since the nuclear tragedy of Hiroshima and Nagasaki happened so long ago, it has become an abstraction without image and concrete expression, which does not give a real perception of the impending catastrophe. That is also the reason that people are unperceptive to the danger of the nuclear weapons race. In this connection, many cannot see and understand the qualitative difference between the effects of conventional weapons and of nuclear ones. Therefore, lack of experience and abstraction is considered by some people to be a possible psychological mechanism hampering the evaluation of the reality of nuclear war and its aftermath.

The more real the threat, the more serious are the behavioural reactions in response to the threat.

Today, the public has reached such a level of understanding of the real threat to people's security, which has resulted from the quantitative and qualitative build-up of nuclear weapons, that the protest movement against the nuclear threat has become more active. I think that in the coming years, public protest and the actions of different sections of the public against the further slipping towards nuclear conflict will grow, and will be more effective in preventing nuclear catastrophe.

The threat of devastating nuclear war hanging over the people of the world, military hysteria, creates the situation of mass stress, emotional pressure, intellectual disharmony and nervous overload. As a result of the arms race, people are not sure of their tomorrow, and they are afraid of the future. This increases the incidence of cardiovascular, oncological, psychic, and other diseases.

We must also not forget that nuclear weapons have a 'memory' — a long radioactive memory. And we doctors know this better than anyone else. Those who survive a nuclear explosion will be confronted by the danger of the fatal consequences of leukaemia or malignant tumours for the rest of their lives. Hundreds of thousands of people will live in fear of cancer and the transmission of the genetic defects.

The fatal aftermath of nuclear war is connected not only with the direct effect on people of the destructive factors of nuclear weapons, but also by the emergence in the region, or even over the whole planet, of an environment unsuitable for life. Changes in chemical composition and physical properties will happen to all elements of the biosphere: air, rivers, seas and oceans, plant and animal kingdoms, as well as electro- and magnetospheres.

Further, conditions of life and health will be considerably influenced by the destruction of industrial enterprises. Production of goods vital for life (food, clothes, medicines, etc.) will cease, or some greater or lesser part of them will be destroyed, and it will be impossible to receive aid from outside.

Destruction of industrial plants, especially of chemical and related branches of industry, with their depots of raw materials and ready goods, accumulations of liquid and solid industrial waste (often highly toxic), will lead to the discharge into the atmosphere of large masses of toxic products to contaminate the atmospheric air, water in open reservoirs, underground water and soil. The destruction of sewage reservoirs may lead to a sharp increase of acids in waters, fish and plankton. Mercury, which very often can be found in sediments, is transformed by acids into a highly toxic substance. A 'delayed-action' chemical 'bomb' will be actually placed under the soil. Thus, massive instant ejection of toxic substances into the environment would lead to mass poisoning of the survivors.

Today, it is becoming increasingly obvious that the madness of the arms race demands an exorbitant price from humanity without bringing it security, and at the same time is depriving it of the possibility of solving the most urgent problems of socio-economic development.

Indeed, vast expenditures on the technology of nuclear war are being made at a time when millions of peoples are starving, are suffering from all kinds of diseases and are being denied elementary material benefits. While a mere four dollars are earmarked for each minute of research devoted to the prevention of heart attacks from which four persons die in the world every 60 seconds, 250 000 times more money is allotted for the development and production of the instruments of murder.

Specialists maintain that 60 per cent of the total sum of military expenditure in a year is enough to build 600 000 schools for 400 million pupils, 30 000 hospitals with 18 million beds, or 50 million comfortable flats for 300 million people.

We understand that we are speaking about horrible things, some even tend to accuse us of frightening people, but the truth that concerns everyone is better spoken aloud and bluntly.

Albert Einstein used to say that 'we shall require a substantially new manner of thinking if mankind is to survive'. This manner can only be based on truth, courage and honesty. It is these criteria that have brought together tens of thousands of honest physicians in our movement, dedicated to the cause of preserving life and health in various parts of the world regardless of their nationality, and their religious and political views.

The first congress of the International Physicians for the Prevention of Nuclear War was held in 1981 in the United States. The second congress was held in Britain a year later, when 32 countries were represented. The seeds we sowed two years ago are now bearing fruit. We are witnessing the growth of an anti-missile and anti-nuclear movement all over the world.

The Soviet Committee, 'Physicians for the Prevention of

Nuclear War', has done much to inform large sections of the Soviet and world public on the question of the medical consequences of nuclear war. It published a book entitled *The Danger of Nuclear War* (Chazov, Ilyin and Guskova, 1982), presenting the Soviet physicians' viewpoint. Soviet television broadcasts, in which the consequences of a nuclear war were discussed by Soviet and American scientists, were seen by about 150 million viewers. These questions are widely discussed in the Soviet press, on radio, and in special medical periodicals.

Soviet physicians consider themselves a part of the growing international movement of scientists and doctors who sincerely and selflessly serve one goal — to do everything possible to avert the madness of nuclear war and to defend the health and life of all people on earth.

The Soviet Committee was actively involved in preparations for the Third International Congress, 'Illusion of Survival in a Nuclear War', which was held in the Netherlands in June 1983. The participants have considered some organisational issues for a future World Congress of Physicians. Its scientific programme and problems are to be discussed at working group sessions. One of the main reports at a plenary meeting of the Congress will be made by Soviet scientists on the immediate biomedical impact of a nuclear war.

We firmly believe that there is no fatal inevitability of nuclear war. We want to prevent the world from becoming engaged in a new round of the arms race and we are convinced that the Soviet Union and the western countries, especially in Europe, are equally interested in this. The main thing now is to halt the process of sliding into a new world war and to prevent the situation from becoming unmanageable. This is a cause to be fought for and our contribution to this struggle is to tell people honestly what awaits mankind if a thermonuclear war is unleashed. Those who entrust us with their health and life, our patients, must come to realise the reality of the nuclear danger.

None of us can make a great contribution to the cause of peace by working single-handed. Only by joining our forces and broadening our movement shall we be able to wage an effective struggle against the nuclear weapons race and make our contribution to the cause of preserving life on earth.

Time does not wait. Too much explosive material has been accumulated in the world. There is no place for nuclear weapons on the earth, and the sooner they are destroyed, the more reliably can we save people from the dangers which aggression poses to mankind.

We believe that reason will triumph.

Reference

Chazov, Y. I., Ilyin, L. A. and Guskova, A. K. (1982), 'The danger of nuclear war: Soviet physicians point of view'. Novosti Press Agency Publishing House, Moscow.

12 A SOLDIER'S REPORT
David Hackworth

David Hackworth

Nuclear weapons — or deterrence, the non-use of nuclear weapons — are the backbone of the United States' defence policy, and by irrational extension, the backbone of its foreign policy. Diplomatic means of resolving conflict are, for most purposes, extinct. The reliance on the military solution — in both East and West — has brought us to the point now where America and its alliance partners live in a state of undeclared war with the Soviet Union. The escalation in war talk and war toys cannot go on forever . . . and when it ends, so will we.

Common sense shouts that the military solution does not work anymore. Atomic warheads have changed the rules of war, and have made global warfare obsolete. If man does not abolish the concept of using the bomb, then the bomb will abolish humankind.

My credentials are simply that I know war. For the past four decades, I have followed, in the words of Macarthur, the sounds of muskets. I fought in five 'hot' wars: WWII, Korea, Laos, Cambodia, and Vietnam; and was a participant in almost every critical campaign of the first twenty-five years of the Cold War. I know the sound of war . . . and today, the sound is deafening.

I speak as a military man, who spent most of his life dealing in military solutions. I know first-hand the people in the US who are saying that nuclear war is 'winnable, survivable, and manageable'. I recognise the mentality of the generals who are saying that a nuclear war can be limited to a European tactical battlefield or extended into a protracted global strategic war, complete with the traditional winners and losers. And I know well the men like General Haig whose NATO plan called for nuclear 'shots across the bow' in the event of a Warsaw Pact confrontation.

I quit the Army and my country because of the madness of the military solution in Vietnam: the constant escalation, the lies and deceits, and the refusal of the military to even *try* a fresh, lateral approach to the problems of that bad war. I argued with General Haig that the invasion of Cambodia was wrong. I challenged my superiors about the conduct of the war, because few even tried to understand its nature. I see the same people who were telling us that there was light at the end of the Vietnam tunnel, now saying and doing the same thing in Central America. General Wallace Nutting, Commander of US Forces in Latin America, said recently, with regard to America's war in Central America, that 'if we make the evident commitment, without limit . . .' (That means *a blank cheque,* like Westmoreland got in Vietnam.) — we will convince the guerrillas 'they can't win'. He went on to say, 'I can't say at this moment that 150, 200, 500, or *1000* trainers in El Salvador is sufficient.' Like the 100th rerun of a bad late night movie reeling into the shadows, I see Vietnam all over again.

Yet, the military man's rationale cannot be impugned or degraded. He is defending his country, a grave responsibility. His view is narrow, conservative, blinkered and cautious. Like a poker player who thinks he could win if he only had more chips, the military man thinks that if he only had more guns, more soldiers, more warheads — victory would certainly be his. The fact that Vietnam ruptured this theory is not the subject of this discussion, except in so far as the United States seems to have learned nothing from that disastrous example of the military solution.

The military tail is wagging the dog of the American state. America has become a militarised nation. Many key positions, in days past chaired by civilians who thought in terms other than military solutions, are now filled by generals: Haig came from defence contractor, United Technology, to become Secretary of State, and after his 'I'm in charge' performance and America's bizarre behaviour in Lebanon, went back to United Technology. General Scowcroft, who chaired the MX committee, recommended that the Air Force MX system be pursued. This 'unbiased' opinion was not particularly surprising, considering that Scowcroft is a retired Air Force lieutenant general. General Richard Stillwell, Westmoreland's former Vietnam Chief of Staff, is now heading up a top secret spy agency, unknown even to the CIA until they were compromised recently in the Laotian 'Bo Gritz' affair. And there are hundreds of other senior officers in key positions, not only in the Departments of Defence and State, but also spread throughout the 'big business' defence industry. US Senator Proxmire's study revealed that there are thousands of former military top brass floating back and forth between the Pentagon and the arms merchants, each sharing perceived views and ingrained conclusions which emphasise the military solution. As the arms race increases, the generals have more troops, more officers, more room at the top, more toys and boys and stars and bars . . . and more military solutions.

Military men are not all bad. But most go along to get along, for that is the only way to survive and excel in the US military today. After a time, these soldiers begin to believe their rhetoric; they wallow in their fantasies of future glory. But this just does not work when the world stands with one foot on a nuclear roller-skate and the other on a militarized banana peel.

Since 1945, we have had 20 occasions when the bomb was recommended by US generals as the military solution: during international flashpoints, such as Trieste, Berlin, and Cuba, and in hot war situations like Korea, Mat Sui, Dien Bien Phu, and later Vietnam when the escalation got out of control and the military did not know how else to win. I was on the ground during most of these grim confrontations, and know how close the world stood on the brink of destruction.

I grew up with the military leaders who are in charge today. We cut our teeth on the bomb. As a young officer, along with Haig, Vessey, Meyer, and Nutting, I learned how to use nuclear weapons and was as eager as the rest to use them. They were just another weapon. The axiom then was to 'nuke the pukes', and we couldn't wait. I commanded nuke weapons in Germany, and had the Cold War turned hot, to accomplish my mission, or to save my unit . . . I would have used them. As Monty said, 'I'd shoot first and ask questions later.'

Now, these guys are in charge of the whole shooting match, and they are still saying 'nuke the pukes'. They are advising President Reagan — whose own military experience is limited to playing captains in old WWII movies — and he believes them! 'Two bars,' Mr Reagan is fond of saying of his celluloid rank, 'can't argue with four stars.'

The military solution, in both East and West, reached its pinnacle with the insane strategy of deterrence. Like the United States' 70 billion dollar MX missile project, which was so ill-conceived that the experts cannot even agree on how to base it, deterrence has not worked. It has only gotten bigger and badder over the last 40 years, culminating now with the same experts flitting between the strategy of Mutually Assured Destruction and Ronald Reagan's Star Wars scenario, designed to deter the Soviets from outer space, proving that the principle of deterrence has no *earthly* value . . . so why not give it a go in outer space?

The most brilliant discussion of the myth of the deterrence strategy is in the April 1983 *Bulletin of the Atomic Scientists,* in an article entitled 'Common Sense and Nuclear Peace'. And I quote:

> We should not act as if we understand the value of nuclear weapons as a deterrent. The consequences of using nuclear weapons are so out of line with any conceivable scenarios, short of an outrageous military action or the actual use of nuclear weapons, that nuclear weapons may be for most purposes no deterrent at all. Does anyone really believe that an attempt to unify the two Germanys justifies the destruction of both? Did the existence of nuclear weapons keep the Soviets out of Afghanistan, or the Americans out of Vietnam? Nuclear weapons may have contributed to deterring a head-on collision between the superpowers, but the risks we take by relying solely on nuclear weapons to inhibit lesser U.S.-Soviet confrontations are too great and are therefore unacceptable.

So much for deterrence . . . on to diplomacy:

> The very triumph of scientific annihilation has destroyed the possibility of war being a medium of *practical* settlement of international differences. No longer is war the weapon of adventure whereby a short-cut to international power and wealth can be gained. If you lose you are annihilated. If you win, you stand only to lose. War contains the germs of double suicide.

So said the biggest hawk of them all, General Douglas Macarthur. He, like most of the senior military leaders from the WWII era . . . Eisenhower, Montgomery, Mountbatten, and Bradley . . . all went on record as saying that nuclear warfare demands that governments find another way to resolve conflict. What frightens me is that my former colleagues haven't gotten the word. They have allowed the nuclear sledgehammer, like mustard gas in 1916, to become the ultimate military solution, to the exclusion of diplomacy and statesmanship.

Because of the militarisation of both the US and the USSR, the military establishment is not only in charge of preparing for war — or deterrence, which is the same thing — but more and more in charge of making peace. It is like putting known criminals in charge of the police department.

For example, in Geneva, the man heading up the US disarmament delegation is named Edward Rowny. He is a lieutenant general. Rowny once wrote that it was impossible to negotiate with the Russians because of the 'Mongol influence'. The Russians, he said, were descended from the Mongols and respected nothing but brute force. They would never negotiate an arms-control agreement unless compelled to by confrontation with US military superiority. General Rowny, on whose shoulders the future of the world rests, is going through the motions of arms control while really spoiling for a fight.

I see the same ever-increasing momentum that created the insanity of the Vietnam War during the Johnson and Nixon Administrations — with mad talk of winning, prevailing, light at the end of the tunnel-ing — in the Reagan administration with its increasingly irrational policies and war talk. Here is a quick review of the Washington War Rhetoric:

Eugene Rostow, former presidential advisor on Disarmament: 'We are living in a pre-war, and not a post-war period.'

Rostow's assistant said: 'It is possible for any society to survive a nuclear war. Nuclear war is a destructive thing, but still in large part a physics problem.' (Whatever that means!)

Richard Pipes, until recently a member of Reagan's National Security Council: 'War with the Soviet Union is inevitable.' And, 'The US has undertaken a campaign aimed at internal reform in the Soviet Union and shrinkage of the Soviet Empire.' And then he added: 'Soviet leaders would have to choose between peacefully changing the Communist system . . . or going to war.' And later:

'Nuclear war would be like an amputation; traumatic, but not necessarily fatal.'

Colin Gray, who works with General Rowny at Geneva, has said: 'Victory or defeat in nuclear war is possible.' And later he said: 'The US should plan to defeat the Soviet Union and to do so at a cost that would not prohibit US recovery.' He went on to say that the losses in a nuclear war in the US could be kept at a 'compatible level of 20 million dead'.

Louis O. Giuffrida, head of Reagan's civil defence programme, said: 'Nuke war would be a terrible mess, but it would not be unmanageable.'

Aren't these comments insane? But there's more . . .

Giuffrida's assistant, when asked if democracy would survive an all-out nuclear war with the Soviet Union, said: 'I think they would eventually, yeah. As I say, the ants eventually build another ant hill.'

For perhaps one of the more bizarre statements of them all, one turns to the President's assistant, Edwin Meese. 'Nuclear war is unadvisable.'

George Bush, the Vice President, explains the military solution with characteristic insight: 'You have the capability that inflicts more damage on the opposition than it inflicts on you. That's the way you can have a winner.'

And finally, the President: 'We could survive a nuclear war. It would be a survival of some of our people and some of our facilities, but you could start again. It would not be anything that I think in our society you would consider acceptable, but then, we have different regard for human life than those monsters do.'

No wonder the Russians are scared. No wonder America's allies are scared. Helmut Schmidt, former Chancellor of West Germany, recently said that he doubts the Reagan administration is serious in its negotiations in Geneva. He said he would not endorse further deployment of nuclear weapons in Europe until the Administration showed evidence of good faith in these negotiations. He said that there had 'never been greater neglect for European participation in the alliance than under Carter and Reagan'.

And alliances are what makes Australia the meat in the nuclear sandwich.

I recently wrote a letter to the Prime Minister, and received a reply from a senior minister in his stead. This minister informed me that our Government is opposed to mass nuclear destruction, and is working for arms control and disarmament. He told me that 'the relationship and alliance we have with the United States is a fundamental feature of Australia's defence and security arrangements . . . and that the Australia/US joint facilities make a positive contribution to deterrence which reduces the likelihood of nuclear war, and consequently reduce the risks to Australia'. He told me that 'successive Australian governments have judged that our best interests lie in remaining part of the Western alliance'. And he said that the Government is committed to work toward 'meaningful detente and effective measures of arms control using, inter alia, Australia's influence as an ally of the United States'.

Let's briefly explore these myths which threaten the security — the very being — of this great nation.

First. We've already discussed the myth of deterrence. It does not work. It simply delays the inevitable. Therefore, *there is nothing* Australia can do to make a 'positive contribution' towards it. It does not reduce the likelihood of nuclear war; instead only postpones it while making bigger, badder bombs which will do the deed many times over when Richard Pipes and the Reagan Administration's 'inevitable war' finally comes.

Second. Government policy opposes the use of all techniques of mass destruction. But the US/Australia joint facilities are the systems which control, command, and support the weapons of mass destruction. These facilities are the eyes of the West. They provide intelligence data on America's enemy, the Soviet Union, and command elements of America's nuclear strike force. Each day the deterrence spring is wound tighter. Now it's 'launch on warning', and Australia provides the warning. Dr Geoffrey Smith, in a recent article, entitled 'Terror Australis', submits that 'by harbouring key US bases, Australia has become, in fact, little different from a nuclear power.'

Third. Dr Smith submits as well, that Australia, by keeping these US bases, 'has accepted the risks of nuclear war.' There is no doubt as to the truth of this statement. In war, you first strike to blind your opponent; accordingly, Australia, which, I repeat, provides the eyes of the American nuclear giant, will be among the first struck. These bases, *in Australia,* are *high-priority Soviet targets.* Despite what Prime Minister Hawke has gone on record as saying — 'The advantages outweigh the risks.' — the joint facilities *do not* reduce the risks of nuclear war for Australia.

Four. Should Australia consider its relationship with the United

States as a fundamental part of its defence and security arrangements? Do Australia's best interests lie in remaining part of the Western Alliance, as 'successive Australian governments' have judged? As a military man, I can see no advantages. America is now a one-ocean navy, and that ocean is certianly not the Pacific, because America gets its oil from the East. And in the nuclear missile age, we have but eight minutes before the first Soviet warhead impacts on Australian soil. No alliance, no treaty partner, no deterrence or Star Wars strategy would have a chance to stop those millions of tons of radiated TNT from striking the Lucky Country.

Five. Can Australia work towards meaningful detente and arms control using its influence as an ally of the United States? Helmut Schmidt spoke for Europe . . . if America ignores European participation, why would they consider Australia's more valid? They don't even notify us of their many nuclear alerts. Just one example was on 9 November 1979. On that day, America was three minutes from launch because of a computer error. Mr Fraser was not informed, or Mrs Thatcher, to name just two close alliance partners. And meaningful detente between Australia and the Soviet Union is virtually impossible if Australia continues to propagate the 'us versus them' mentality through its alliance with the United States.

We need to change our mindsets immediately. We must give our ingrained habits and assumptions electric shock treatment. We need a fresh way of resolving our differences. Not in the context of the 1930s appeasement, but in the realism of the 1980s. We have to forget 'us versus them', and accept 'you and me, warts and all' in the Atomic Age.

Every day, people are becoming more and more concerned, and they turn to their elected members for support, but the Government seems so unwilling to discuss this issue in public.

The people are confused. Experts claim there are too many bombs. Other experts claim that there are not enough. Each side sells its product like the successful insurance man hustles his family protection plan. The terminology is mind-numbing. The threat is mind-numbing. The fear is mind-numbing. And because the Government remains silent *when there is so much it could do,* many people just tune out.

What can the Government do? First, it must realise that the nuclear issue transcends politics, and all parties must unite to find new solutions to old problems. We must forget the 1942 Battle of Midway and fight the 1983 battle to prevent this country from being destroyed by the insanity of a superpower war, on which both sides are spending themselves broke, neither dares to fight, but fight one day, they will . . . through an accident, a miscalculation, or the act of a madman.

Australia must wrench itself out of the stranglehold of its one-sided treaties with the United States. We must stand on our own feet, and we must take a fresh, *moral* look at the issue. We must explore new, perhaps even radical ideas and options in the name of arms control. For example, why could not Pine Gap and the other US secret facilities become, instead of the eyes of the American military, the eyes *of the world,* on the world? The technology is there; the facilities could be joint US/USSR verification centres operating for peace under the auspices of the United Nations. Would this not be the ultimate Australian peacekeeping force? Here, on our own soil, yet for *the entire world?*

Patrick White has said, 'Besides this global issue, what else matters?' And he's right. All other problems — economics, ecology, elections, or exposés — all are dwarfed by the shadow of nuclear war, which grows longer with the passing of each day.

The world is crying out for peace. There is no reason why the the Lucky Country should become one of the first irradiated dominoes in the superpower game. Australia can *and must* light the way for world disarmament and lead us out of this nuclear nightmare. If Australia took the first step, both East and West might be brought together to stop the madness and save the world.

And we have not a moment to lose.

13 ECOLOGY AND PEACE: SOME EXPERIENCES OF THE GREEN PARTY

Roland Vogt

Roland Vogt

In 1976 when Petra Kelly returned to Germany after a lecture tour of Australia, she was able to give the anti-nuclear movement in Germany some very encouraging information. She told us of the Green Ban against the transportation of uranium, the 'Keep Uranium in the Ground' Movement, and the co-operation between the Anti-Uranium Movement, some trade unions and Aboriginals.

We needed such encouragement at that time because we had just witnessed a very traumatic demonstration in favour of nuclear energy, organised by an influential trade union in co-operation with a pro-nuclear public relations agency. The Australian example greatly inspired many of us, including Nina Gladitz who made the documentary film *The Uranium Belongs to the Rainbow-Serpent,* which was well received by the Anti-Nuclear Movement in West Germany and even shown on television.

I now hope to be able to re-invest some of the encouragement your movement gave to us seven years ago.

The history of the Anti-Nuclear Movement, including the breakthrough of the Green Party, is connected with the inter-linkage of the ecological and the peace movements which together form the life movement, or the Green movement. This inter-linkage did not just happen, it was prepared intellectually and organised practically.

A new era of enlightenment?

If the slogan of the era of enlightenment, following the philosopher Immanual Kant, was '*sapere aude*' (dare to use your own brain), the formula today has to go beyond *sapere aude*. We must be courageous enough to understand the inter-connection of problems and dare to act following our understanding of how things are inter-connected. The first attempts to develop a public understanding of the inter-connection and common action of the anti-nuclear branches against nuclear power plants and nuclear weapons were made in 1975 and 1976 — and they failed.

In 1975, for example, I tried to convince some regional citizen action groups struggling against the nuclear power plant in Wyhl to organise an Easter March against both nuclear power plants and nuclear weapons. Intimidated by the propaganda strategy of the state government, some leaders of the regional group did not dare to organise the Easter March in that form. The Easter Marches of the anti-nuclear weapons movement of the sixties had been labelled 'leftist' and in 1975 the local population which resisted the construction of a nuclear power plant in Wyhl was afraid of being called Communist. Consequently, they organised an international meeting against nuclear power plants and neglected to oppose nuclear weapons. I took part in that imperfect demonstration and made the acquaintance of Petra Kelly, who came there as a speaker, and that was the beginning of a deep personal friendship.

More than one year later, the newborn political couple and the future co-founders of the Green Party, Kelly and Vogt, made an address to the public in the same region, proclaiming that the resistance against nuclear power plants had to be a Peace Movement as well. When I said this in an open-air speech some citizen shouted, 'That's political, that's Communist thinking!' The reaction of the public was split but the majority voted in favour of my continuing the speech.

At that time we used to argue that it was not logical to fight against a dozen nuclear power plants without mentioning the 6000 atomic warheads on German ground. We tried to prove that the so-called civilian and the military use of nuclear fission were Siamese twins, based on the same insane substance, uranium. We said that with the export of the atomic fuel cycle to countries like Argentina, Pakistan or Israel, the original producers actually exported the desire to build the atomic bomb into neighbouring countries like Brazil, India and Iraq. The availability of the nuclear cycle in the hands of a neighbour nurtured the fear that he could make the atomic bomb first. Some years later, when Israel bombed a nuclear reactor in Iraq, we did not have to prove any more in an academic way, that this argument was not altogether wrong. *And,* we could argue with the Australian example, that the fight against atomic weapons requires blocking the nuclear cycle at its source.

In some publications of the ecological movement we used to introduce a special section called 'Ecology and Peace'. So over the years we prepared, step by step, a growing awareness of the inter-connection between ecology and peace. It was recognised then, that ecology means peaceful relations both between humankind and nature and also among human beings themselves. And finally we could easily introduce into the peace manifesto of the Green Party, this basic paragraph:

> The life-style and mode of production which are based on a never-ending flow of raw materials which are then used in a wasteful fashion, provide the motivation for the use of force to acquire those raw materials from others. In contrast the responsible use of raw materials in an ecologically based life style and economy reduces the danger of the political use of force on our behalf. Ecological politics within the community create the conditions for the reduction of tension and the opportunity for peace in the world.

Action-planning 'ecology and peace'

A more practical way to bring about the co-operation between the ecological and the peace movement was to organise common action-conferences. So, we achieved some joint demonstrations in local areas where nuclear power plants were to be built close to existing military installations.

New approach to disarmament

The first conference of that kind took place in October 1979 and thus some peace groups and ecological citizen action groups were able to react in a *common statement* against the proposed decision to deploy new medium-range missiles on Western European soil. We proclaimed instead of that so-called 're-armament' a new approach to disarmament. We called it 'first, well calculated steps towards disarmament'.

We recognised that we are compelled to disarm and that we need new approaches to achieve real disarmament. Never in history have negotiations ever resulted in disarmament. We do not trust the mutual balanced disarmament concept. Instead of this, we require real disarmament measures. The first calculated steps must be taken, not under the condition that the other side takes the same step (that has never worked), but in the *expectation* that our action is the first step and that it will be followed by a step from the other side and so on. In this way the doves on both sides encourage each other and take the initiative in what we call the *dis*armament race.

Two branches on one tree

With the help of such concepts and with common actions we brought together two branches: the ecological movement and the new peace movement — the joint Green Movement as a life-oriented movement. If 'freedom' is regarded as the top value of the West and of the conservative, liberal and so-called Christian parties, if on the other hand 'equality' is the top value of the East and the Communist and socialist parties, then 'life' is the utmost highest value of the Green Movement. The Green Party is a party within movements such as the ecological, the peace, the third world and the women's movements.

Block-free: de-block the brains

The Green Party supports the position of European Nuclear Disarmament (END), first launched in the United Kingdom and supporting block-free politics and nuclear free zones in Europe. Block-free politics demands first of all that we de-block our brains — in Germany not an easy task after decades of brainwashing, anti-Communist propaganda and so on. To de-block brains requires an autonomous position, to decide questions based on our own needs, and not following the formula that what is detrimental to my enemy is good for me, or that the mere fact that my neighbour possesses something means that I must get the same thing. This argument is often used in Germany to justify the nuclear industry (East Germany, Russia and France have got it — not to have it means that we will be inferior in competition).

Our main concern at the moment, to stop the next armament circle by preventing the deployment of Pershing-II and cruise missiles, does not mean that we can neglect the other mass destructive weapons. We criticise them also from the point of view of the final consumers of those weapons, that means the future victims whether they be victims of the SS-20, the French missiles Pluton-I, -II, Hades, the British Trident-I, -II, or the so-called tactical US atomic weapons, as they are all targeted to explode on German ground anyhow. Our slogan for that block-free idea is 'Neither to the East nor to the West but loyal to one another'.

Five members of the Green Party, among them four members of Parliament (including Petra Kelly and myself) recently have stressed that position by demonstrating in East Berlin with the banner of the East German autonomous Peace Movement 'Swords to Ploughshares'. That it was understood as a non-violent rather than a hostile action was proved in a letter from Chairman Honecker to a message we had left in the headquarters of the State Police in East Berlin. During my temporary arrest in the German Democratic Republic, I had some interesting disarmament talks with different civil servants. My proposition is that we should all have such private disarmament talks continuously, with everyone. We need them in our countries because up to now, the official disarmament talks have failed. Private disarmament talks are probably more substantial.

Social Defence: necessity is the mother of invention

Another alternative which fits well with our ecological block-free and non-violent orientation is the Social Defence idea. Following the old saying that necessity is the mother of invention, from time to time people have invented non-military means of struggle. This could happen when their political leaders did not accept military action to resist an invader (1968 CSSR), or when the army was too weak to be used to fight back military intervention from outside. Such an experience happened in the twenties, when Germany resisted the Belgian-French invasion of the Ruhr region (*Ruhrkampf*).

The Green Party is a non-violent organisation within the peace movement and prefers the conversion from military to non-military defence to any military concept. In our party it is considered a proven point that security cannot be achieved by military means in the international threat system of the atomic age. From the ideological point of view, one might add that military apparatus and supply agencies share the fate of all human institutions aiming at absolute security. They produce the opposite — insecurity, force, pathological learning. On one hand, if the worst comes to the worst, the result is the destruction of human beings and nature, and on the other hand, if circumstances are favourable, the end result is the break-up of the institution which is founded on faulty construction.

Like human beings, institutions are subject to error. The destructive capacity of the military systems combined with error equals extermination (of mankind probably), whereas Social Defence endures human error. Military concepts encourage the military pattern of conflict resolution. Civilian concepts set free the productive imagination of every citizen concerned.

I was happy to learn that Brian Martin and the Canberra Peacemakers share our views about the practicability of Social Defence. They are doing good work in introducing this idea into the political process of communities in Australia. So it is not my job to 'export' that idea to Australia.

The European Peace Movement

As a member of the European Peace Movement at this moment, I feel more like a beggar than someone who can offer something. We urgently need assistance.

(a) *Governmental help*

I feel that the European and the US Peace Movements alone have not the capacity to bring about the necessary swing to disarmament. But in order to prevent a nuclear holocaust our generation must ensure that real disarmament measures are enforced. We have to wrest disarmament from the political and economic forces in power, including Dr Kohl, Mrs Thatcher, Mr Reagan and Chairmen Andropov and Honecker. The Australian and Canadian Governments should help the many people in our countries who desire disarmament. A first joint action could be an international Summit for Survival co-chaired by the Australian Government as suggested by Stella Cornelius.

(b) *Help from the Australian communities*

In the coming months and at the latest by next fall, we have to withstand the attempt to deploy the US Pershing-II and cruise missiles. Parts of our peace movement, such as the women of Greenham Common and non-violent blockaders in Western Germany, have to face enormous problems, including financing their actions and paying fines.

The peace movement in Comiso, Sicily, needs an important amount of money (Aust. $50 000) to buy land just opposite the deployment site to establish an International Peace Camp there. One hundred and twelve US cruise missiles in the south of Sicily, straight opposite the North African coast, will obviously not be directed exclusively towards the Warsaw Pact countries.

My proposition is that the Australian Peace Movement sends help instantly to the European Peace Movement at this critical time — not only idealistically but also financially. The British community could then help the women of Greenham Common, the Italian community could make some contribution to our friends in Comiso, and the German community could send funds to the affinity groups that organise non-violent blockades of all the proposed sites of Pershing-II and cruise missiles.

Appeal

I implore the Government and the people of Australia — do not repeat the tragic mistakes we made in Europe. Do not allow your continent to be maltreated by one of the superpowers as a military colony. Do not accept that targets are installed. Even under Australian control they would be targets; controlled targets remain targets. Do not allow yourselves to be hostages to the superpowers. We, in Europe, are such hostages and perhaps it is too late to free ourselves from that condition.

'. . . act globally as well'

One of the slogans of Green Parties in Europe says, 'Think globally — act locally'. That is what we have tried to do in recent years and that is what created the ground for our first successes. But now we have to learn to act locally and globally, to act together.

Let this be the message. Let this be the beginning of an interconnected Australian-European block-free peace movement.

I have confronted you with some strange and unthinkable options. To avoid 'the unthinkable' — the extermination of mankind — we are charged to re-invent the world.

Or let me say this in the words of Erich Fried, a German poet, living in London: 'Who wants the world to remain as it is, doesn't want the world to remain'.

14 THE ROLE OF THE AUSTRALIAN GOVERNMENT
Susan Ryan

Susan Ryan

Despite the peripheral rôle of foreign policy issues in the 1983 electoral campaign, Labor's theme of reconciliation and reconstruction is good news for international as well as national issues. We hope and intend that our aims of consensus, of the resolution of conflict will find support in our region as they did in the electorate.

I must clarify what we mean by consensus. We were not — we are not — foolish enough to suggest that the lion will now lie down with the lamb. Strongly held differences between groups with different interests, different perspectives, different beliefs will continue, as it should in a free country. What we offered in 1983 was the hope that these differences could be managed and resolved with some civilised respect for the legitimacy of the rights and interests of opposing groups. We argued that as Australians, despite our pluralism, we could on some essential matters have common values and objectives.

This belief in the need for conflict to be resolved peacefully with respect for the differing stand points of others is being carried over into our foreign policy. It is a belief which will be difficult to sustain in the face of the reality of the present state of the world. Today, savagery and violence between nations is commonplace. The abuse of the most basic human rights is not an aberration but is practised widely as a conscious act of policy by many governments and terrorist groups.

Poverty and oppression are the lot of the majority of the world's population. We see interminable struggle in Africa and the Middle East. We see violent revolutions sweeping Latin America after centuries of dictatorship. We see Kampuchea and Afghanistan occupied by foreign armies. We see a Europe which remains an armed camp. Most ominously of all, we see that the flower of detente between the two great powers has withered, and the tensions between the two growing each day.

Confronted with this reality, it would be easy to succumb to cynicism and a sense of helplessness. What part can a small power like Australia play in a world in which such giant military, economic and ideological forces clash? We could be tempted to withdraw within ourselves and worry about nothing but putting our own economy right, and perhaps about our immediate region.

Certainly the new government's foreign policy gives priority to our own region. The Minister for Foreign Affairs, Bill Hayden, made his first two overseas visits to the five countries of ASEAN and to Papua New Guinea. Prime Minister Bob Hawke's first official international visits will be to Papua New Guinea and Indonesia. These facts are not accidental. Nor is it an accident that this Government has taken the considerable risk of offering to do what it can to facilitate moves towards the settlement of the Kampuchean issue. The prospects of success are not high, but our willingness to try demonstrates our commitment to a more stable and peaceful region.

Our Government is a democratic socialist government; that means that within Australia we are committed to the redistribution of wealth and power so that every person can live at a reasonable standard of material security, and has the freedom to fulfil his or her human capabilities. Modest though they sound, these goals cannot be achieved without major changes that are difficult and controversial. However, we are unashamedly idealist in our aims, if practical in our methods. We reject power without purpose — the occupancy of the Treasury benches for its own sake. For us the proper exercise of power is to move Australia towards equality and security.

Even if by some miracle we were able to achieve these aims at home immediately, that achievement would itself be meaningless if nuclear catastrophe were to befall our world.

We cannot pretend that this catastrophe is an irrelevance to Australia just because we are a vast island continent remote from the likely centre of a nuclear exchange. No country can escape the consequences of nuclear war simply by an accident of geography. Nuclear war would involve us all. Moreover, Australia has on its territory joint Australian/US facilities which are related to the nuclear balance. The North West Cape communications base is a prime nuclear target, because of its crucial role in command control and communications functions with US nuclear-armed submarines. Nurrungar and Pine Gap have critical roles in verification processes associated with nuclear arms control. The Government supports the presence of these facilities, on certain strict conditions, as long as they continue to support a system of nuclear deterrence.

For all these reasons, the new Labor Government has already announced its determination to play a much more active role in international arms control and disarmament. We do so with a sense of urgency at the growing danger of nuclear war. Relations between the superpowers have not been so bad since the 1950s and early 1960s; but unlike the 1950s and early sixties the nuclear balance is no longer overwhelmingly in favour of one of the superpowers. Each side has a rough equivalence in capability. That means that in a crisis such as occurred over Cuba, one side is now much less likely to back down than, it was twenty years ago.

The existence of so much destructive capability when the superpowers are politically so hostile is in itself bad enough; but an even greater danger is the way that the fragile strategic balance is being undermined by the advance of technology. New, more accurate and powerful weapons are threatening deterrence by suggesting the capability to win nuclear war. Thus we hear terms such as 'flexible response', 'limited nuclear war', 'nuclear superiority', 'counterforce' and 'controlled escalation'. These terms are nonsensical. Weapons, war-fighting, superiority — each presupposes that you are using force to achieve specific purposes, and may emerge bloody but victorious.

However, nuclear weapons are not just a bigger bang — they are not just conventional arms writ large. Once a nuclear exchange starts, it will not be limited, controlled or flexible. It will rapidly become full-scale nuclear war. The list of people who have stated this is compelling: President Brezhnev, former Secretary of State Haig, Lord Mountbatten, US Admiral Gayler, Lord Carver the former UK Chief of Staff, the recently retired head of the US Joint Chiefs of Staff General David Jones, and the former Australian Defence Minister Sir James Killen.

In this environment, while being courageous and creative in our efforts, we must be realistic about what we can achieve. We acknowledge there is something inherently immoral about weapons which are not just totally indiscriminate in their potential for mass destruction of civilian populations, but which, if they are ever used, will certainly destroy all that mankind has so painfully wrought. Yet they do exist. They will continue to exist as long as great powers and the standard bearers of different ideologies continue to believe that others are bent on their destruction. Therefore, even though the only long term aim must be to strive for eventual total nuclear disarmament, while these weapons exist, we cannot throw up our hands in helplessness and hopelessness.

Parallel with our efforts for total disarmament, we must also work for a stable system of deterrence and undertake the grinding and often boring task of controlling new dangers which spring from new technological capabilities. We recognise the limitations of deterrence (it presupposes a degree of rationality in human affairs which we can only dare to hope will last in the long term); but if we are prepared to settle for nothing less than immediate total disarmament — for all or nothing — we are doomed to the noble impotence of passionate rhetoric.

In pursuing our commitment to arms control and disarmament, the Labor Government will not feel obliged to stand aside from the central questions of the strategic balance. Certainly, only the two superpowers can work out the detail of arms limitation agreement between themselves; but we, as host to important US facilities, as an ally of the United States, and as citizens of a world which would be effectively destroyed by a nuclear war between the two superpowers, have the right to put our views forcefully to both. We intend to do so in international forums and in bilateral discussions. Specifically, we will urge that the leadership of both countries should resume control of the technology, rather than allowing technology to become the driving force of our destruction. We will urge them to abandon fundamentally irrational doctrines of limited war and of winning a nuclear contest, which make the likelihood of an escalating nuclear exchange so much greater. We will urge that genuine dialogue between them should occur, with the greatest possible degree of frankness on both sides, and with a readiness to find accommodation and accept restraints. They should work towards an agreed concept of stable mutual deterrence, under which neither side should develop any nuclear capability which threatens to disarm the other. Otherwise, the other side may think it has no choice but to attack first.

Without progress in superpower dialogue and restraint, multilateral efforts towards disarmament have become paralysed. With progress between the superpowers, there is much that can be done multilaterally. The new Labor Government has already taken a number of actions. We have protested vigorously to the French about their nuclear testing in the Pacific. We have stated our opposition to all forms of nuclear testing by all states in all environments, and our commitment to an early conclusion of a comprehensive test-ban treaty. We have stated our intention to raise at the next meeting of the South Pacific Forum a proposal for a South-west Pacific nuclear free zone. We have indicated that in contrast to our predecessors we will not stand in the way of a consensus to hold an international conference on the Indian Ocean as a zone of peace. We have indicated our determination to resist all attempts to undermine the safeguards regime established under the treaty of non-proliferation of nuclear weapons.

We have announced our support for a ban on attacks on civilian nuclear facilities. We are reviewing our record on signature and ratification of, and reservations to, existing disarmament treaties. We recognise that outer space is the coming area of superpower arms competition, and that limits to this development must be set. We will be supporting unequivocally urgent action in the Committee on Disarmament in Geneva to identify issues for negotiation in this area. We recognise that if Australia is to have credibility in attempting to play a more active role in disarmament and arms control, we will have to strengthen greatly our capacity for analysis and policy formulation in this area. The capability of various government departments, notably the Department of Foreign Affairs, to work effectively in the disarmament area is to be considerably strengthened.

We intend to establish a peace research and arms control institute and are currently considering the best methods of achieving this. We have already embarked upon consultations with the Australian

National University for that purpose. The Government has now decided to appoint an Ambassador for Disarmament to represent Australia in all international forums on disarmament and arms control.

Let me turn to extra-parliamentary action. As regards the peace movement, let me say that this Government does not deride the supporters of this movement as the naive dupes of a Soviet disinformation campaign. The role of the peace movement is crucial, because it is making disarmament and arms control an issue of domestic political controversy. I do not think it is unfair to politicians to say that if the argument in favour of disarmament can be presented, not just in terms of rationality and morality, but also in terms of votes, it will be infinitely more persuasive. We are seeing this happening in the United States now over the freeze debate; we are seeing it happen in Europe. Unhappily, it is not happening in the Soviet Union; the Communist Party of the Soviet Union still seems fairly confident of winning any election in the foreseeable future with 99.9 per cent of the vote. The fact that the efforts of people in the Soviet Union to speak out for disarmament, have been so quickly crushed should be condemned throughout the world.

No democracy is perfect but the great advantage, the great justification for democratic political systems is that they allow people to raise their voices in protest and to organise and lobby governments to change their stance on such crucial issues as disarmament. In democratic systems as in Australia, we need both determination by government and pressure from the people if we are to make the best of our opportunity to promote disarmament. One of the most dynamic aspects of the peoples' disarmament movement in this country, as in Europe and the United States, is the involvement of women.

The growing support by women and particularly by women with a feminist analysis for the disarmament movement gives us cause for optimism. There is an obvious connection between the growing political activism of Australian women, their growing support for the progressive policies of Labor, and the increasing numbers of women who are organising to press governments to disarm. Women have historically supported the disarmament movement; all through this century women's groups such as the Women's International League for Peace and Freedom have continued to raise their voices in support of peace in national and international forums. There is an obvious reason for this phenomenon. Women with very few exceptions do not make war, but they are the victims of it. Women lose partners, parents and children through war. They lose their country and lose their lives. Women are raped in war (despite the denials of the RSL), even when there is no war, women lose out by national and international commitments to gigantic arms budgets. Last year, in our own relatively affluent country, the government spent over $4 billion on defence at a time when homelessness, unemployment and poverty are increasingly the lot of women, particularly women with dependant children. So for these reasons it is only to be expected that women see that they have no stake in war and every reason to support peace.

But feminism provides an even more basic attack on militarism than the traditional concerns of women. The ideological basis of the women's movement in Australia, as internationally, is a rejection of the male values of aggression and violence, and an advocacy of such values as co-operation, consensus, tolerance and trust. I believe that a recognition of these positive values of feminism will promote a strong women's disarmament movement; but it is essential to link the revulsion for war with a rejection of the political and social and economic structures that leave women in a powerless position. I believe that many more women will be persuaded to support the disarmament movement if they see that the disarmament movement is not only about persuading governments to disarm but is also about changing the kind of society that produces militarism. Issues of domestic and social violence cannot be separated from the violence of war. The needs of the society to care for and support its vulnerable members, be they children, the aged, or the disadvantaged, cannot be separated from the argument about armaments expenditure. It is important that policy makers in all countries and of course especially here in ours, see the connections between the women's peace movement, the aspirations for a society in which all citizens can live in peace and security, and the negotiations between superpowers.

Finally I repeat our Government has the will to proceed towards the goal of nuclear disarmament, but the people have the responsibility to sustain and support us in this most important of all endeavours.

15 WOMEN AND THE PREVENTION OF NUCLEAR WAR

Nancy Shelley

Nancy Shelley

The nuclear obscenity is the ultimate scientific and technological outcome of a way of thinking which excludes from its view all but what it wants to take into account; a framework which allows the penetration of information which it designates relevant and disdains all other; a mode of thinking which does its own defining of reality and defines out of importance, sometimes out of existence, other ways of viewing the world; a framework of thinking which is, by its nature, narrow, unimaginative and arrogant; a way of thinking which suffers from blindness through that narrowness and arrogance, which has achieved dominance in our society and which carries with it a violence to alternatives.

In our society, it is men who have developed the science and technology of which the ultimate products are nuclear weapons. It is women who have been made invisible by the dominance of male thinking, who sometimes have been defined out of existence, and who are constantly objects of violence.

Women across the world are saying: No more will we allow you to destroy us, our children and our children's children. The very survival of the planet and humanity are at stake this time. We know — and you too would know if you were not blind — that the road you are on, and have dragged us along, is the road to annihilation. We have good cause not to trust your judgement any more.

Of course there are men who are appalled too. The issue is not women against men. The issue is, however, the structure of men's thinking and the dominance of white male culture. I would hope that men, perhaps particularly those who feel impatient, would make an attempt to listen and understand when women speak and act.

Different Perspectives

Berit Ås (1982) writing in Women's Studies International Forum offers the following synopsis of her article:

> Women are victims in all wars. Men plan them, they train for them and they conduct them. They have the capacity to inflict pain and death, destroy peoples and territories. The different roles of men and women in wars dispose the two sexes to different thinking, feeling and action with regard to warfare. A similar difference of interests exists between men and women with respect to the army. For women, their children, most often sons, have made up the armies of history. As cannon fodder, women's work and life intentions are disrespected and destroyed. For men, the army consists of *comrades* who become the most important people for them because their life-and-death chances are bound up with their fellow soldiers. Finally, through history the owners of territories have been men. Women have very seldom held land, property or slaves. More often than not they have been sold, captured, stolen or even given away. These three contrasting relationships towards warmaking, the army, and territories provide perspectives for men and women when they discuss military matters and policies for peace.

As Dorothy Green (1983) has written: 'Even men who are opposed to the nuclear arms race are often fascinated by its technological and political complexities, and by the opportunities it affords for endless intellectual exercises'.

Berit Ås (1982) continues:

> Our demands are for a peace policy. Women demand concrete plans and clear goals for detente and disarmament. They know that the billions spent on the armament race have been stolen from the hands of the hungry and the oppressed people of the world. They regard technological weapon discoveries which increase the ability of the apparatus of violence to kill with greater pain, and which heighten the effects of terror, with disgust and protest.

This difference in perspective is also observable among peace researchers. Female researchers find that there is 'too much mindless weapon counting' and 'too little about the human and social consequences of the arms race.' (Brock-Utne)

Australian women have likewise expressed grave doubts about the proposed structure, operation and priorities of the Australian Peace and Development Research Institute which is being advocated here.

Marcia Yudkin (1982) says: 'War is a male institution', by which she means

> that as a matter of social and historical reality there exist structures, patterns of behaviour, customs, practices, and organizations into which males, as a matter of social and historical reality, have been socialized. In our society, men have been the actual and potential perpetrators of war . . . men have been in the position to prepare war, declare war, and wage war. Any women who have fitted into the mould have been anomalies.

It must be seen and acknowledged that *only* when we perceive these differences, arising as they do historically and socially, and come to respect the differing approaches, can we together participate in both the debate and the task of preventing nuclear war. Men have to refrain from saying (and thinking), 'Yes, I agree with what you say, but . . .'; and the but often goes on with: 'the way to achieve all those things is by the route I propose; (e.g.) we must remove the threat of nuclear war first and we'll attend to other things later.'

Men have to learn that to *demand* the acceptance of their argument is to *require* the acceptance of their framework. That is not working together, it is oppression. There are many factors which must be stated clearly, and understood, if we are to work together in this most vital task.

Experiences of this Century

Some lessons can be learned by considering the events of the last seventy years. The war that was fought between 1914 and 1918 had several names. It was called the First World War, because it involved a large number of nations in the fighting. It was called the Great War, because the human losses were so great. It was called the War to End All Wars, because people were appalled at the slaughter and shocked at the devastation.

Out of that war was born the League of Nations with its hope of bringing people to the conference table. The hopes of humankind were misplaced and we experienced the Second World War. This time more fighting men died, the civilian casualties increased tremendously, constituting 45 per cent of all casualties (the USSR alone suffering twenty million dead), more nations across the world were involved, larger areas of countryside were devastated, and the weapons used increased in horror as that war was conducted on land, sea, and in the air.

Out of World War II was born the United Nations with its Charter and its machinery for resolving conflict. The UN, in its nearly forty years has done some remarkable things — too little of its achievements are known. Yet, today, some one hundred and fifty wars later and after nearly thirty million have died in those wars, we are faced with the build-up of unprecedented terror.

Yet again, the men who operate those structures of negotiation have not come to terms with their underlying fascination with, and respect for, the power of might. The UN is in danger of being as ineffectual as the League of Nations, because we deny it the machinery and authority to act to prevent the bullies of this world acting in contempt of those aspirations which are written into its Charter. We are still captive to the rhetoric which proclaims the principle, the cause, the glory, the requirement of dominance and the myth of victory, while the misery of slaughter and desolation, of hunger, poverty and disease are made conveniently invisible, or the fault of those who suffer them.

It is hard to have confidence that peace is the overwhelming object of the representatives of government who assemble at the UN in New York and who arrange conferences in other parts of the world. We can draw attention to the absence of peace in the titles of conferences to deal with the Environment (Stockholm 1972) and with World Population (Bucharest 1974), yet to the inclusion of it in the title of the UN Conference for Women (Mexico 1975). For nearly a decade now women have been meeting, studying, researching and working together under the theme of 'Equality, Development and Peace'.

The Global Plan of Action arising out of the 1975 Conference states, under the title International Cooperation and Control of International Peace, that:

> In order to attract women more and more into the promotion of international cooperation, one should recognize and encourage women's efforts for *peace* as individuals and in groups, and in national and international organizations, through the expansion of international ties among nations, strengthening of international peace and disarmament, combating colonialism, neocolonialism, foreign domination and oppression, apartheid and racial discrimination (UN, 1975).

We may well ask whether it is hoped that women, more than men, are able to find peace for the world. If that is the case, we should ponder the sincerity of that hope, when as Berit Ås observes: 'While those who are rich in resources, with access to technology, money and political institutions, have been responsible for war, the responsibility for peace seems to be placed on those who lack resources, technology, money and political power' (UN, 1975). Likewise it is salutary to note, as Robin Morgan (1982) writes: 'Women in all cultures have been assigned . . . the positive values of 'humanism', pacificism, nurturance, ecoconsciousness, and reverence for life, while these values have been regarded by Man as amusingly irrelevant.'

Is it any wonder women are now saying: You've had your chances to bring about peace. You've had it for centuries and you haven't managed it. We want to stress the REAL priorities, and these are the priorities of life, not of fear, of co-operation not competition and hollow victory, of human needs not weapons construction and bargaining — for now, it is with the whole world in the balance that you dare to bargain!

The Nuclear Age and Morality

Out of the ashes of World War II came the beginning of a new era — an era which has its roots in death and destruction. The names

Alamogordo, Hiroshima, Nagasaki, Bikini, Enewetak, Maralinga, Mururoa, Nevada Desert, Windscale, Three Mile Island, send shivers in our memories — gravestones replacing milestones on a road whose end is holocaust.

The words of Yoko Ohta (1963) express this as she describes what she witnessed in Hiroshima:

> Each of us had for a time done everything possible, without knowing for sure what exactly it was that we were doing. Then we awoke, and now we wished to speak no more. Even the sheep-dogs that roamed about ceased to bark. The trees, the plants, all that lived seemed numb, without movement or colour. Hiroshima did not somehow resemble a city destroyed by war, but rather a fragment of a world that was ending: Mankind had destroyed itself, and the survivors now felt as though they were suicides who had failed.

Japanese women tell that when a woman in Hiroshima is pregnant she receives no congratulations. People wait in silence for nine months until the child is born to see if it is all right.

The Nuclear Age has spewed other names into our thinking: superbomb (Hiroshima is too small!), Minuteman, Polaris, SS-4, SS-5, Poseidon, neutron bomb, SS-20, Pershing-I, Pershing-II, cruise, Trident, MX; and acronyms ABMs, ASMs, ASW, SRAMs, IRBMs, SAMs, MRBMs, SLBMs, ICBMs, MIRVs . . . We use these words, making sentences which contain them. We make statements like: 'Each Trident missile has 17 independently targeted warheads. Each Trident submarine contains 24 of these missiles. That means 408 cities can be targeted by one submarine. The fire power of each submarine is equivalent to 8 World War IIs. The US plans to build 30 Trident submarines and there is one of them already in the Pacific.'

We use these words, we make these sentences, when every fibre of our being should be shouting: we are talking the ultimate of obscenities. Instead, we deny our humanity still further by pretending a rationality in this insanity, and become co-opted into the enormity as into a conspiracy, by proposing 'rational' arguments to refute those put forward by the madmen who exercise power in the world, and by suggesting 'rational' actions to meet the menace.

In World War I, approximately 95 per cent of casualties were soldiers. In World War II, 45 per cent of casualties were civilians. The prediction for a war using weapons of mass destruction is that civilian casualties could amount to 95 per cent, while the well-trained armies will be able to protect themselves.

We can exclaim with moral outrage, we can recoil with repugnance, yet what do we know about the 'value systems and morality that provide legitimation for the military-industrial complex'? Women scholars, working in disarmament-related fields, have this as a major concern in their research.

Irrationality

Birgit Brock-Utne, in 'The Relationship of Feminism to Peace and Peace Education' (unpublished), recalls a letter from Alfred Nobel, inventor of dynamite, to Berthe von Suttner, the very active peace campaigner and pacificist writer. (It was through her constant pleadings that Nobel donated the money for the Peace Prize.) Nobel maintained he did more to prevent war by his inventing dynamite than she did through all her peace conferences and campaigns. He wrote: 'When the day arrives that two armed nations can destroy each other within minutes, then all nations will shrink back and will dismiss their troops.' That day has arrived, but the troops have not been dismissed. Women endorse what Birgit Brock-Utne writes, that 'his thinking seems naive, irrational and illogical. We find that if the aim is complete disarmament, one must first stop rearmament and start disarming. We do not believe in preparing for war if you want peace.'

All of us are aware of other manifestations of the illogical and irrational, like the increasing insecurity that is the result of ever-greater expenditure upon what militarists call security. Brock-Utne continues by asking the question: 'How is it possible that this highly illogical and irrational thinking be stamped as rational by those in power?' Her answer is:

> those in power also have the power to define the world. They have the power to define concepts, to allocate prestigious words to their own thinking and to stamp the thinking of others through words which have a negative connotation. Those in power also have the power to define concepts like defence, security, justice. They have the power to stamp their own

> thinking as rational . . . even though, according to most criteria, it is highly irrational.

Alongside this, we are all familiar with the tendency of the military to assume almost exclusive use — certainly the dominant use — of words like courage, honour, loyalty and pride. Women will point to how, in a patriarchal society, these virtues, along with rationality and logic, are stamped as 'manly' virtues, whereas to be 'womanly' is used in derision or abuse. We are familiar too with the denigration, and often open violence, that accompanies attacks on those who would define these concepts otherwise. The most significant of these in Australia in recent times has been the reaction of the male community and the military to the marches of the Women Against Rape at Anzac Day Parades. Those women seek to remember all women raped in all wars — thus introducing a part of the remembrance of war which is not to be countenanced. The women's action brings such strong reaction because it touches the nerve of so much of the latent militarism within our society. It dares to reveal something of the accompanying aspects of war — things which are constantly dismissed as unimportant — and it lays bare the contempt with which women are held.

Sexism and Militarism

Betty Reardon, in 'Militarism and Sexism: Influences on Education for War' (unpublished), states: 'Militarism and sexism complement, reinforce and help to perpetuate each other. They also constitute major obstacles to overcoming war, largely as they affect the education and formation of generation after generation . . .' She refers to the desirability of a fixed order which is articulated by the current ultra-conservative and authoritarian movements, and their insistence upon specific unchanging roles — in a society which they intend to keep unchanged. She continues:

> Among these specific roles are those determined by sex. And among the expectations of fixed sex roles, is the likelihood of young men serving in war, war being viewed, as are sex roles, as the inevitable consequence of human nature. Thus, militarism has been a significant aspect of the socialization and education of boys and largely determines what comprises socially desirable

> masculine attributes. These attributes tend to be qualities deemed necessary for military service, such as bravery, aggressiveness, endurance, discipline and the repression of the 'softer' human sentiments . . . [Boys] are also reassured, perhaps as an inducement to be willing, if necessary, to make 'the ultimate sacrifice', that this identity is superior to that of the 'other', the feminine identity.

Betty Reardon traces some of the features of education in American schools and their effects upon children of both sexes. Continuing, she writes:

> If the children's day involves athletics . . . they are urged to 'lay low' or 'waste' their opponents. The powerful lesson of sports in general is that the highest human value and virtue is winning and . . . proving less value and virtue in the loser or the 'other'. If you may some day be called upon to kill 'others', you need to believe in their lack of value and virtue. Small wonder that boys grow up believing that true masculine identity resides in soldiering, fighting and winning, and that some people (especially women who cannot even engage in, much less triumph in the most physically engaging sports) are less valuable than others.

Bee Pooley (1981) draws attention to our conditioning into roles based on gender, saying: 'Thus women are taught to be feminine, the chief characteristic being passivity, and men are taught to be masculine, the chief characteristic being aggression. The oppression of women is also maintained through women believing in and internalising the values of the group in power.'

Although these writers are looking at the US and the UK, similar attitudes are observable in Australia.

Thus in early socialisation and education, in any patriarchal society, men are encouraged, along with the development of attributes which receive acclaim, to believe in their own superiority and to accept the relegation of women to lower levels of consideration. The predisposition, therefore, is there, before the militarists have their full sway. It is easily visible, that: 'Those on the right, who relentlessly work for hierarchical orderings, authoritarian controls and increased armaments, correctly perceive that sexual inequality is the cornerstone of the system they seek to impose on us all, and therefore also work relentlessly *against* sexual equality' (Eisler, 1982).

These two pillars can be observed in President Reagan's policies over the last two years. Accompanying his strident advocacy for more massive armaments, has been a most vocal attack on feminist aspirations. One of the chief exponents of attack on women's rights has been George Gilder, whose book *Wealth and Poverty* President Reagan gave to all his cabinet appointees.

It is no accident that the man who wrote: 'Men should be trained for war and women for the recreation of the warrior', was Hitler's ideologist, Nietzsche. It is clear that the more militarist a society is, the more sexist it tends to be.

There are many aspects of military training which act to desensitise men, to lower those attitudes of humane response to other human beings — in particular, to those who are designated 'the enemy'. Much of this is well known and deplored. What is less acknowledged is the part played in the dehumanising process of relegating women to be objects for men's use. Any resident of Perth will testify to the open supplying of women for the US navy during their frequent use of Fremantle port facilities.

The training of soldiers in masculinist ideology of necessity reduces women in esteem. Some of the training carries with it blatant contempt for women. Sara Ruddick (1983) makes the point that: 'Many marching chants include degrading reference to anything feminine. Sexist terms for women and their bodily parts are common in military discipline as well as in play.'

Berit Ås (1982) makes the point that throughout all war literature the dominant picture of the soldiers is as a husband or lover, and of the women is as wives or whores. Yet, beyond doubt, the majority of adult women in a country will experience the army as made up of sons.

Violence Against Women

Another manifestation of the low esteem with which women are held in a patriarchal society is the toleration and promotion of pornography. The violence done to women in this 'industry' is heinous. No longer is it permissible to discount such practice as fringe activity, nor to dismiss it as irrelevant. Women know this to be an integral part of the militarisation of our society, and fear the increase in the level of violence which has now become 'acceptable' in Australia, as in other countries. If the amount of domestic violence and the amount of pornography are increasing, and as women are the objects of such violence, and if little is done to deal with it, women may well ask those men who hold power in this country why they can ignore that violence and its implications.

There are two points to be made here. First, women link the increase in violence toward them with a growing attitude within our society which will tolerate war, i.e. as part of a process which brings citizens to accept that total violence when the 'leaders' of the world decide the time is right for them to initiate it. If the peace movement, therefore, ignores this, failing to see the symptoms of a malaise, it will be inept in its attempts to prevent war, and in particular, nuclear war. Second: men who are genuinely desiring to stop the madness will be less capable of bringing about their intentions if they fail to understand the extent to which they are predisposed to violence by the socialisation they have been subjected to.

It is totally absurd, and structurally impossible to try to bring about world peace when aggression and conquest are considered synonymous with manliness and masculinity. It is unlikely that nuclear war can be prevented if negotiations are left to men who are necessarily driven by such perceptions of what it means to be manly and those perceptions being so interlocked with the doctrine of 'might is right'.

We have, of course, a few women in the world who have taken upon themselves these attitudes and who see their 'success' as stemming from out-doing men in war-like actions. You will be familiar with the compliments such women receive likening them to strong men, which sustains my point concerning the values of a patriarchal society, and should heighten your awareness to the difficulties men face when they want to be otherwise. In contrast, Elise Boulding's study (1981) has exploded the myth that there are 'no qualified women available' in the disarmament and security field. Yet these women are seldom used. She asks why they are not used. The best known internationally of such women are Alva Myrdal and Inga Thorsson from Sweden, Mary Kaldor from the UK, and Betty Goetz Lall from the US. These women tend to be the ones most frequently called upon, but there are many more women who are well qualified in disarmament issues, yet they are not used.

Olive Schreiner (1983), (who so perceptively in her novels linked women's struggle with the race struggle, and identified the campaign for the liberation of women with the campaign against war), writing from South Africa in 1911, said: 'War will pass when intellectual culture and activity have made possible to the female an equal share in the control and governance of modern national life . . .'

Women are not seeking equal access to the structures of a patriarchal society, to be co-opted into the violence of that society. Women are saying loud and clear: It has to change.

Violence Against the Poor

The second aspect of war that the women's action on Anzac Day dared to reveal concerns its unpleasant side. As remembrance of wars concentrates upon the sacrifice of men's lives and their courage, refusing to acknowledge the squalor, the destitution, the suffering of both soldiers and civilians, so too, the rhetoric accompanying the urging by the military and politicians of support for massive arms expenditure omits to link the effect of such spending upon rising hunger, poverty and disease all over the world. Ruth Leger Sivard (1977) describes it.

> There is in this balance of global priorities an alarming air of unreality. It suggests two worlds operating independently of each other. The military world, which seems to dominate the power structure, has first call on money and other resources, creates and gets the most advanced technology, and is seemingly out of touch with those threats to the social order which have nothing to do with weapons. The other world, the reality around us . . . is a global community whose members are increasingly dependent upon one another for scarce resources, clean air and water, mutual survival. Its basic problems are too real, too complex, for military solutions.

Terrorism is one of the words which has been adopted for very specific application. Yet is it not terrorism when 12 million babies die before their first birthday? — when 450 million people suffer hunger and malnutrition? — when malaria could be eradicated with the allocation of half of one day of the annual military expenditure in the world? — when instruments and techniques of torture are exported to military dictatorships to suppress ordinary people in their efforts to obtain food and better living conditions? — when in the interests of the superpowers, the wishes of whole populations, as in Belau and Poland, are ignored, because of global strategies for possible military confrontation? Surely, the greatest act of terrorism today is the nuclear arms race, which is responsible for this reality of what is happening throughout the world. Guns take food from the hungry. Women are saying that unless we link our action for preventing war with the total scene of the violence done to the poor throughout the world, we are being hypocritical and will fail in our intent. Political action is needed to solve political problems, and the allocation of resources — financial, material and human — is done by political decision.

Feminists draw attention to the violence which permeates the world and make the connections between rape and militarism, poverty and the arms race, linking destruction of the environment, torture of dissenters and intervention, with pornography, pollution, competing to win and cruise missiles. (Can one maintain that the 30 whales which beached themselves to die on the Pagoda beach in 1980 have no place in our thoughts and actions for preventing nuclear war, when we know that ten years earlier the US army had dumped thousands of canisters of nerve gas 100 feet down in the Atlantic, ten miles off the Pagoda beach? The canisters had, of course, begun to leak.)

As the women who took action at the Pentagon in November 1980 declared: 'All is connectedness'. Effective action to prevent nuclear war, therefore, MUST include simultaneous action to deal with human needs, actions to protect the environment, and action to eliminate all forms of violence.

Violence to the Spirit

There are other forms of violence which need to be spoken of: the violence of despair, of resignation, of powerlessness, and the violence of the lie. It has been said that war begins with the acquisition of human bodies, but long before that happens we are subjected to the silencing of human souls. It is this silencing which has taken place over the last forty years, which rose to great heights in the cold war of the fifties, and is reaching greater heights today. If that violence to the human spirit has been complete, then we could say confidently that all our efforts to prevent nuclear war would be in vain. It would mean that people had been sedated into acceptance of their annihilation, and no amount of exhortation would counter that final blasphemy: the militarists' security of anaesthesia. However, in the words of the feminist song:

You can't kill the spirit
She is like a mountain
Old and strong
She goes on and on and on . . . (Women's Peace Camp, 1983)

The large numbers of people who have come out on to the streets in the cities of the world, bear testimony to the fact that the human spirit is still active, alert and unwilling to be conned by the violence that has been done to it. This is exemplified in the poem Gillian Booth read to the magistrate when she was arrested at Greenham Common. This is part of it:

. . . what do you do with someone like me
the animal called human who . . .
flies screaming in the face of official logic
unrepentantly and happily dissident
to join her friends who were occupying that sentry box
at the entrance to this monster
that all my life has breached my peace
what do you do when I admit
that I did nothing wrong
and tell you that after two men got hold of me
and dragged me back to the gate
I ran to the side gate laughing
slid the latch
and ran right in again
and that the only way I can be stopped
is to silence me by death . . . (Women's Peace Camp, 1983)

It is in the action of ordinary people that we can put our trust, for basic to that action is their awakening, their determination, their anger and their compassion.

In Britain, in the last two years, national membership of CND alone has jumped from 3000 to 50 000. Here in Australia, three years ago in Brisbane the Peace Rally attracted 80 people. This year there were 6000 and the city was made a nuclear-free zone three days earlier.

In no way, of course, have we yet done enough. However, in our clamour for particular action to prevent nuclear war, we must remember two things. First, there are still millions of people who need to be awakened to the enormity of nuclear terror, and those who have brought us to this brink of catastrophe have much money, power and conniving skills at their disposal to attempt to counter our work. (Let us not be ignorant of the promises of material prosperity which blind people to the growing dangers of authoritarian power over them.) Second, there is no possibilty that we shall succeed unless we arouse the awareness of those millions more, and in that task we must respect the tenderness and frailty of their awakening, for they will not change one form of violence for another.

This is clear from one woman's description of her withdrawal for a time from the Peace Movement. From Sicily she wrote:

> It was as if everything for which I've struggled . . . was being taken over by the 'male world' and they are calling me back into line. (Women's Peace Camp, 1983)

That woman has rejoined for she can now articulate her position:

> Our revolution is our refusal to accept the powerful, the 'official' politicians, everything male that does not represent us, not even when it disguises itself as the fear of nuclear death and sometimes speaks a language 'similar' to our own but in reality is only well camouflaged . . . (Women's Peace Camp, 1983)

It is essential we work at preventing the polarisation of society — for that can only aid militarism.

Women and Non-Violence

The actions of women in the Peace Movement in different parts of the world give a stimulating picture. Women are contributing so much which passes without notice in the media, is often derided and certainly is unsung. Women have developed networks around the world, and not only is the information flowing, but the spirit of women is growing, meeting each of us in our particular part of the globe, touching us each in our own lives, strengthening us and renewing our creativity as we come to celebrate life again in all that we do.

Women moving for peace are so serious about their commitment that they are able to bring joy into their action. It is a life-affirming response to the death-proclaiming nuclear threat. It is hope which defies crippling despair. It is empowerment which overturns powerlessness.

One woman's view of non-violence is provided in the account of the Greenham Common Story (Women's Peace Camp, 1983). She writes:

> To me, non-violence is not just about the way we as women approach direct action. We cannot 'fight' violence with violence; the ends and the means must be the same to achieve Peace without Bloodshed. So many times groups . . . involved in political and social change have resorted to violence, usually in frustration. This time we must get it right. If it isn't right this time we may not get another chance.
>
> In our everyday life, conflict in some form or another is inevitable. The way in which our society is structured dictates the conflicts and frustration within it. The way we, and in turn governments, resolve and respond to conflict is crucial: we need new tools to deal with situations. For centuries the way in which men have participated willingly in ritual blood-letting ceremonies called wars are now unacceptable. Military generals, governments and out-dated laws — which permit and encourage the acceptability of death, killing and war — must end. 'Winning' and 'losing' have lost their meaning on a world where nuclear weapons are capable of 'overkill'. As women, we have remained silenced for too long, silenced by the argument of defence — the balance has been tipped and at last we see the ultimate insult to our intellect: 'nuclear weapons'.
>
> Women are no longer giving their silent consent to wars. Generation after generation of sons and daughters have, and are still being, killed, innocently and willingly.
>
> As women we have the power to change the next generation. We will need every scrap of strength and determination to stop the rot of indifference and apathy that still encourages little boys to play with war toys and believe that 'might is right'.
>
> Angry? Of course we are angry, like the earth if she could speak in words would tell us.
>
> But our anger is compassionate — still. We do not want to kill because of it. We are all just realizing that we have the responsibility for the Earth and our species' survival and because of this women from all over the world are saying NO MORE — WE HAVE HAD ENOUGH.

> *Non-violent direct action*
>
> The non-violent direct action that we use shows us many things. It shows us we do not need 'superhuman' courage or physical strength, but intelligence in the true meaning of the word. It shows authorities/governments that we no longer have respect for their barbaric values and ways of resolving conflict, and it also shows us that as women we can be effective when we trust each other and our survival instinct. Very often 'symbolic' actions have been sneered at because they do not achieve complete immobility or destruction of 'property'. I think to be effective we must take responsibility and make our anger symbolic — if we rid ourselves of symbolism we certainly risk violence and potentially the very thing we were aiming to replace.
>
> 'Non-violent' or 'Pacifist' men sometimes ask me — 'Why is it *Women's* non-violent direct action?' My answer is usually that women can be more effective, that we cannot be provoked as easily as men, and that the military and the police cannot remain credible if publicly they used mass violence on women and children who were being non-violent.
>
> These are not the only reasons. Women have more of a vested interest in Peace. Our children need a future! As women we know what oppression is — and we know what society offers us in return for co-operation.
>
> Many women say 'I cannot be active because I have other commitments' — this is the way our society keeps us quiet. We cannot protest too loudly because of repercussions and conflicts in our personal lives. We appear to be damned if we do *or* don't. As women we must say NO MORE. WE HAVE HAD ENOUGH. WE HAVE NO CHOICE ANYMORE.

There is then a commitment to non-violence because of a commitment to bringing about a non-violent world. There is a belief in non-violence as showing a necessary way: an alternative to violence. There is belief in the power of non-violence. There are those who doubt the power of non-violent action. In this respect it is important to understand the twin elements which are necessary: on the one hand, condemnation of certain actions and practices; on the other, respect for the person engaged in that action. Barbara Deming (1983) sums it up this way:

> It is precisely solicitude for his person in combination with a stubborn interference with his actions that can give us a very special degree of control . . . We put upon him two pressures — the pressure of our defiance of him and the pressure of our respect for his life — and it happens that in combination these two pressures are effective.

However, non-violence is not simply a technique of struggle; it arises from a belief that all life is worth love and respect. The evil of a system has to be attacked, not the person who operates that system.

Women's Action for Peace

Women's ingenuity is evident in the variety of action they have been engaged in in many different countries of the world.

The story of the Shibokusa women of Kita Fuji is heroic, desperate, and amusing. After centuries of working the very poor land at the foot of Mount Fuji, under the right of common people to cultivate it, the people of Shibokusa were displaced by the military in 1936. The Japanese military were replaced by the US army, and then, after their withdrawal, the land was kept by the Japanese government. After staging a protest in 1955, the people had some land given back to them, only to have it reclaimed after the trees they had planted were fully grown. When the men were forced to leave the area in search of work, the women took over the struggle. They built a series of cottages around the military base and began to live in them. In 1970, one thousand riot police turned up to evict them, and they faced the possibility of death. On that occasion they surrendered, so that they could continue the struggle.

> The Shibokusa women make it their business to disrupt military exercises. In groups of up to ten, they make their way into the exercise area (there are a host of routes, they say, the secret of which they keep to themselves), crawling around the undergrowth and popping up in the middle of the firing. They plant scarecrows to decoy the troops. Sometimes they'll build a fire and sit round it singing and clapping their hands, totally ignoring officials who try to move them on. They are frequently arrested and taken to the police station. 'They are quite gentle,

> because they are afraid of provoking us — they hate it when we start screaming, and the police have realized that though we are physically easier to arrest than men, we're more trouble afterwards! Men put up a fight, but once it's over they just give everything away. We never give our name, age or anything. We just say we're so old, we can't remember when we were born or who we are . . . (Caldecott, 1983).

Most of these women are in their late fifties or sixties. They know they are part of the wider anti-war movement. For them, they experience militarism as violence against the land. They say: 'Mount Fuji is the symbol of Japan. If they are preparing on her flanks, how can they say Japan desires peace?' (Caldecott).

This kind of long-term resistance is the first of its kind in Japan.

Scandinavian women have arranged women's marches: from Copenhagen to Paris over several weeks in the summer of 1981 and to Moscow and Minsk in 1982. The skill with which they negotiated the manner of their demonstration and their passage with the Soviet Union, at the same time standing firm with their own basic requirements, has much to offer to professional negotiators!

Two women who came to the UN Special Session on Disarmament in New York were from Israel. One was a Jew, the other an Arab. They were inseparable in New York. In Israel, they work quietly and incessantly to bring people to know each other and to break down the barriers between their people.

Three hundred Australian women took part in a cavalcade to Canberra in February 1983. After a rally at the War Memorial they went to, and entered, Parliament House, 'tabling a number of Bills in the Senate'. The women then stood on the lawns outside Parliament House, forming a huge circle, linking hands to project loving energy and support to all women all over the world engaged in the struggle for peace.

There are the women of the Argentine who walk and mourn regularly for the 'disappeared'; the Dutch women building tremendous grass-roots involvement of women across Holland; the six women of West Germany, formulating an appeal and collecting 70 000 signatures of women opposed to the manufacture and stockpiling of nuclear weapons, and promising to practise civil disobedience.

In 1980, several thousand American women went to Washington and took part in action centred on the Pentagon (King, 1983). They began by moving through Arlington Cemetery and there commenced a mourning ceremony, 'walking through the centuries of carnage'. Walking slowly to the sounds of a drum beat, and with moaning, they continued mourning until they reached the Pentagon. They planted gravestones commemorating women who had died, radiation victims, victims of rape and racism and war. They wept without difficulty. Their action was sufficient to bring employees of the war department to the windows, even to the steps of the building, to look.

Next the women began to express their rage and chanted: 'WE WON'T TAKE IT'; 'NO MORE WAR'; 'TAKE THE TOYS AWAY FROM THE BOYS'. The crescendo built up, and the Pentagon employees 'looked astonished, some laughed and some looked very uncomfortable'. Next they began to encircle the Pentagon and to sing as they joined arms, rejoicing when they realised that there were enough of them to complete the encirclement. Then while some entrances were blocked, at one, women wove the doors shut with a web of woollen weaving. That action is one that has had a profound effect. Many women were arrested and some were imprisoned.

What must not be underestimated is the empowering that happens for each woman who takes part in such action. Nor should it be ignored that those who go about their 'work' without thinking of the consequences of their employment — as in the Pentagon, as with many soldiers — are challenged.

There are the women of Comiso demanding that cruise missiles will not be placed in Sicily. Agata Ruscica (1983) wrote: 'What pushes me is that this imaginary thread, which some Sicilian women are building with the women at Greenham . . . continues to exist and other women strengthen it.'

Indeed, Greenham is a source of energy and inspiration on a number of levels. The tenacity of the women involved; the strength of their determination; the creative inventiveness of their continuing activities; the cheerful defiance in the face of arrest, imprisonment, harassment and derision; their endurance of two British winters; their conviction and ever-growing intention to continue their protest until they are sure no missiles will be deployed there — all are strands in a strengthening thread.

So much of what we have taken for granted in our society, and the way in which the nonsense has crept up on us, make it abundantly clear we have to turn it on its head — expose the irrationality, challenge the law-abiders and enforcers in what they are upholding, and, above all, increase people's awareness at all times and in all places.

The women of Greenham have done this to great effect: talking with women prisoners they met when they themselves were imprisoned and conscientizing many of them, confronting the politicians at Westminster, and the judges and magistrates in court. The most common charge they have had to answer is that of 'breaching the peace'. Katrina Howse (1983) had this to say to the magistrates:

> What are you doing to keep the peace? The power you are using is supporting nuclear weapons. It supports binding women's voices, binding our minds and bodies in prison so our voices cannot be heard. So our warning of death is being repressed. But we cannot be silenced. And I cannot be bound over. I AM ASKING YOU TO KEEP THE PEACE. WE ARE NOT ON TRIAL . . . YOU ARE.

Australian women are part of this worldwide awakening and solidarity of women. The networks are growing rapidly and strengthening. Already, women are strong in all groups in Australia working for peace.

White Australian women have begun to meet with Aboriginal women and Islander women, learning from each other how the nuclear madness affects us in differing ways. There are links with women from the Pacific Islands — Belau, Papua New Guinea, Fiji, Hawaii, Tahiti — in their struggle for a nuclear-free Pacific. There are contacts too with women in Asia: women in Japan, the Philippines, Thailand, Indonesia, Korea, Sri Lanka, and India.

Indians traditionally utter the word for peace three times at the beginning of every sacramental rite: 'Shanti, shanti, shanti'. That triple utterance signifies that the power of peace is generated if the will, the means and the end are each of them grounded in peace. Likewise, in order to overcome militarism, we require an equal portioning of our energies to changing fear, injustice and oppression into hope, equity and liberation. If we would prevent nuclear war, we must, at all levels of our life be life-affirming. Our policies must arise from the *will* to enhance the well-being of all humans, they must work at the *means* to fulfil human needs, and they must have human welfare as their *goal*.

In the Australian context this will mean working to prevent the mining and export of uranium, that base of the nuclear chain, that source of death by radiation, that destroyer of Aboriginal lands. It will mean working for a nuclear-free Australia, Pacific and Southern Hemisphere, where people's livelihood is protected, their source of food uncontaminated, their way of life respected. It will mean repudiation of what Helen Caldicott has called 'our suicide pact with the US'. It will mean our withdrawal from global military strategies. It will mean working for the removal of the US military installations on our soil and our military presence in Asia. It will mean ending the violence done to women, to the poor, and to the spirit. It will mean strengthening and developing methods of conflict resolution which respect human beings and their differing interests and needs. It will mean welcoming and valuing diversity and encouraging creativity. It will mean educating our children for peace.

Many women around the world are no longer abashed at the size of the stand we must take in face of the obscenity of the nuclear arms race. They are undaunted at the action their knowledge of life leads them into taking. Their feet are grounded in the reality of the real world with its suffering, poverty, and violence. Their morality stems not from absolutes, but from compassion. Their impetus is not blood-letting, but life. Their strength is based on trust, their affirmation is truth. Their goal is a non-violent world.

> The winds of Time are changing now
> As woman grasps each Woman's hand
> Encircling earth, embracing LIFE
> Weaving webs of hope across the Land. (O'Connor, 1983)

Confidence; complete faith; It is urgent. You can't expect to be led, you must take personal responsibility (Johnson, 1983).

Look long and deeply into the blue of an autumn sky and feel the gentleness and fragility of the eternal as the balmy air fills your lungs. Stand in the golden light of a tree as it catches and gives out the sun's glow, and respond to the warmth and its ripeness. Listen to the full-throated song of a magpie, and wonder at the power of sound to move you to rejoicing. Hold the eye of another with your own in a moment of unspoken unity, and know the exquisite tenderness within you. Take a very young baby into your hands, and experience the miracle and the vulnerability of human life.

If we ponder these miniatures and distil from them their verities, we may come to understand what is involved as we set ourselves the task of preventing nuclear war. Certainly, unless we keep them in view, together with that delicate quality we experience, even in their recall, we shall be in danger of losing our way in a labyrinth of death-proclaiming doctrine whose purpose and end are devastation and destruction. Our task, in contrast, if we are to prevent nuclear war, is to be life-affirming, and the affirmation of life carries with it: robustness in the face of wrongheadedness, violence and evil, tenderness in the presence of fragility, and joy in the business of living. Shanti, shanti, shanti.

References

Ås, B. (1983), 'A Materialistic View of Men's and Women's Attitudes to War', *Women's Studies International Forum, 5,* pp.355-60.

Boulding, E. (1981), Womens International Feminist Quarterly, quoted in Brock-Utne, B., *The Role of Women in Peace Reserach*. Unpublished, pp.3-4.

Brock-Utne, B., 'The Relationship of Feminism to Peace and Peace Education'. Unpublished, p.2.

Caldecott, L. (1983), 'At the Foot of the Mountain. The Shibokusa Women of Kita Fuji' in Jones, L. (ed.), *Keeping the Peace. A Women's Peace Handbook 1,* The Women's Press, pp.104-5.

Deming, B. (1983), 'Revolution and Equilibrium' quoted in Ruddick, S. (1983), 'Preservative Love and Military Destruction: Some Reflections on Mothering and Peace'. Unpublished, p.37.

Eisler, R. (1982) quoted in Brock-Utne, B., 'The Relationship of Feminism to Peace and Peace Education'. Unpublished, p.2.

Green, D. (1983), 'Curbing the Men of War'. Book review of *Keeping the Peace. A Women's Peace Handbook* in 'The Age' newspaper, Melbourne, 30 March, p.19.

Howse, K. (1983), *Women's Peace Camp.*

Johnson, R. (1983), reported in 'Sunday Times' newspaper, 10 April.

King, Y. (1983), 'All is Connectedness: Scenes from The Women's Pentagon Action USA' in Jones, L. (ed.), *Keeping the Peace. A Women's Peace Handbook 1,* The Women's Press, pp.46-81.

Morgan, R. (1982), 'A Quantum Leap in Feminist Theory in *Ms Magazine,* December, p.101.

Nobel, A. Quoted in Brock-Utne, B., 'The Relationship of Feminism to Peace and Peace Education'. Unpublished, p.4.

O'Connor, J. (1983), 'Women on the Winds of Time'. Unpublished poem.

Ohta, Y. (1963), Quoted in Jungk, R., *Children of the Ashes,* translated by Constantine Fitzgibbon, Penguin.

Pooley, B. (1981), 'Feminism and the Peace Movement', *Peace News,* 11 December, p.16.

Reardon, B., 'Militarism and Sexism: Influences on Education for War'. Unpublished, pp.2-3.

Ruddick, S. (1983), 'Preservative Love and Military Destruction: Some Reflections on Mothering and Peace'. Unpublished, p.37.

Ruscica, A. (1983), Letter from Comiso in *Women's Peace Camp.*

Schreiner, O. (1911), 'Women and Labour 1911', quoted in Jones, L. (ed.), *Keeping the Peace. A Women's Peace Handbook 1,* The Women's Press.

Sivard, R. (1977), 'World Military and Social Expenditure, 1977', quoted in Murphy, D. (1981), *Race to the Finish? The Nuclear Stakes,* John Murray.

UN, Glabal Plan of Action (1975) quoted in Ås, B. (1983), 'A Materialistic View of Men's and Women's Attitudes to War', *Women's Studies International Forum,* *5*, pp.355-60.

Women's Peace Camp (1983).

Yudkin, M. (1982), 'Reflections of Woolf's Three Guineas', *Women's Studies International Forum, 5,* p.263.

16 THE ROLE OF THE AUSTRALIAN CITIZEN IN A NUCLEAR WAR

Patrick White

Patrick White

As an epigraph I'd like to quote from a poem by the Australian poet Robert Gray, *To the master Dōgen Zenji* (who lived from AD 1200-1253)

He said, All that's important.
is the ordinary things.

Making the fire
to boil some bathwater, pounding rice, pulling the weeds
and knocking dirt from their roots,

or pouring tea — those blown scarves,
a moment, more beautiful than the drapery
in paintings by a Master.

— 'It is this world of the *dharmas,*
(the atoms)
which is the Diamond.'

For those who may be mystified, *dharma* is the Buddhist truth, the Hindu moral law; again the atoms are those small, ordinary things, as well as the truth, the Diamond being the acme of pure Truth.

I must say I groaned when invited to speak at the conference — like the friends and neighbours I bail up and ask, 'Are you going to take part in the anti-nuclear march?' In the last couple of years I've been doing this sort of thing constantly, often repeating myself, becoming an avoidable Doomsday bore. But anything of importance — like a garden, a human relationship, a child, a religious faith, even the most convinced brand of atheism has to be worked on constantly if it is to survive. So I agreed to speak this time round, and am starting off again to try to explain to my fellow Australians how to prepare themselves to face nuclear war. On this occasion my attempt is made far more difficult in that I am an amateur surrounded by experts in the sciences. But here goes. We are all in it together, and I expect many of my ordinary fellow Australians are as ignorant as I am of the developments of technology and the seemingly endless varieties of nuclear weapons.

My particular concern is how we may develop the *moral* strength, not so much to face as to call off the nuclear war with which the world is threatened. I feel it all starts with the question of identity. In recent years we have been served up a lot of claptrap about the need for a *national* identity. We have been urged to sing imbecile jingles, flex our muscles like the sportsmen from telly commercials, and display a hearty optimism totally unconvincing because so superficial and unnatural. Those who preach this doctrine are usually the kind of chauvinist who is preparing his country, not to avert war, but to engage in it. Anyhow, this is not the way to cultivate an Australian identity. For one thing, we are still in the melting pot, a rich but not yet blended stew of disparate nationalities. And most of us who were transplanted here generations ago, either willingly or unwillingly, the white overlords and their slave whites, are still too uncertain *in ourselves.* Australia will never acquire a national identity until enough *individual* Australians acquire identities of their own. It is a question of spiritual values and must come from within before it can convince and influence others. Then, when our individual identities, united in one aim, cluster together like a swarm of instinctively productive bees — as opposed to that other, coldly scientific, molecular cluster — we may succeed in achieving positive results.

But how to discover this personal identity? I'm always hearing remarks like, 'I feel insecure, I have no confidence . . .' When I tell them that I who have had everything one can expect from life in the way of recognition, awards, money and so forth, feel only intermittently secure and confident, many of those who hear me believe I am putting on an act, while others, who had considered I am one who surely knows the answers, are depressed to find that, by my own admission, I don't. What I do know for certain is that what is regarded as success in a rational, materialistic society only impresses superficial minds. It amounts to nothing and will not help us rout the destructive forces threatening us today. What may be our salvation is the discovery of the identity hidden deep in any one of us, and which may be found in even the most desperate individual, if he cares to search the spiritual womb which contains the embryo of what can be one's personal contribution to truth and life.

We must become aware of what Aldous Huxley refers to in his remarkable book *Ends and Means* as the 'existence of a spiritual reality underlying the phenomenal world and imparting to it whatever value or significance it possesses'. Huxley saw the ethic of non-attachment as the means of attaining awareness of this spiritual reality, because the practice of non-attachment entails the practice of all the virtues — most important charity, but also that of courage, and the cultivation of intelligence, generosity, and disinterestedness.

The ideal of non-attchment has been preached again and again in the course of the last 300 years. It is found in Hinduism, the teachings of Buddha, the doctrine of Lao Tsu, in the philosophy of the Greek Stoics. The Gospel of Jesus is essentially one of non-attachment to the things of this world, and of attachment to God. What the Jewish philosopher Spinoza calls 'blessedness' is simply the state of non-attachment, just as Spinoza's 'human bondage' is the condition of one who identifies himself with his own desires, emotions, and thought processes, or with their objects in the external world.

Again to quote Aldous Huxley, speaking from as far back as 1937, 'Closely associated with the regression in charity is the decline in men's regard for truth. At no period in the world's history has organized lying been practised so shamelessly, or, thanks to modern technology, so efficiently or on so vast a scale as by the political and economic dictators of the present century. Most of this organized lying takes the form of propaganda, inculcating hatred and vanity, and preparing men's minds for war. The principal aim of the liars is the eradication of charitable feelings and behaviour in the sphere of international politics. Technological advance is rapid, but without progress in charity, and awareness of the spiritual undertones and needs of everyday life, it is useless.'

What was true in 1937 is even more pertinent in 1983. If Huxley were alive he might be appalled by his own perspicacity.

But to return to my own experience and the disappointment or disbelief of those who look to me for a lead. When I tell them I don't know the answers, I've got to admit I'm not being strictly truthful. I do, or I *have* known them, and shall again, but only intermittently, the result of a daily wrestling match, and then only by glimmers, as through a veil. None of the great truths can be more than half-grasped. I doubt I should have arrived anywhere near my inklings of them if it weren't for what I sense as links with a supernatural power.

Some of you will see it as a sign of reaction and weakness to introduce mysticism, perhaps even necromancy, into a situation where power politics and increasingly sophisticated technological resources would seem to be leading us inevitably towards nuclear war. However, because I've been asked to give some idea of how I think the Australian people might prepare themselves to face such a situation, I can only stick my neck out and offer my humble beliefs. If I become an outsider by doing so, this won't be a great hardship as I've been that as far back as I can remember — something strange and unacceptable in the eyes of those who believe they see straight. At least it's given me courage of a kind, which I'd like every Australian to acquire. I'd like them to rootle round in their unconscious and find this personal identity, the moral strength which is floating there amongst the trash — the filth. Oh yes, the trash is there in me too, otherwise I shouldn't be able to understand the violence which takes over in such events as bikie riots, the deliberate burning of forests, the destruction of schools — these and many other impulses which would contribute towards the act of violence we fear most.

So let each of us search for the good faith in us which may help save the world, even if we risk turning ourselves into outsiders in this materialistic, muscular Australian society. If we are to give consideration to this momentous issue we must go apart from time to time — apart from our families and friends — and they could think it very peculiar — our occasional evaporations, though we may be skulking only in the next room, or in a trance at the sink — till we find the courage and words to explain the reason for our behaviour — when perhaps we could find ourselves applauded by others who are contemplating similar action. For those who are still afraid of being seen as grotesque supporters of a lost cause, let them take notice of a human chain linked in protest throughout Britain and Europe, gathering resistance in the United States and nearer home the human river which recently flowed through the cynical streets of Sydney and down over the amphitheatre of the Domain, as in other capital cities of Australia. It's obvious we aren't alone.

Still, it is always dispiriting to come across the hordes of unconverted, as I did on my way to the anti-nuclear demo on March 27th — the thousands pouring into the Sydney Show and Sportsgrounds, and to wonder if anyone was aware of a cosmic threat. Did any of them give a passing thought to what might be done about it? Remembering that most human beings are conservative and tend to perform the actions that require least effort, to think the thoughts that are easiest, to feel the emotions that are most commonplace, to give rein to the desires that are most nearly animal, made the dilemma no easier to bear.

Most of these complacent souls have embraced the consumer religion. They have allowed the department store non-culture to persuade them that the accumulation of possessions — cars, TV sets, and unnecessary domestic appliances — is the sum of happiness. The craving for possessions and money, from the humble hire-purchase level, to the smash and grab tactics of the tirelessly acquisitive rich, from the alderman to the union leader and cabinet minister, and finally the dictator of a superpower, has become an epidemic disease. In such a climate, distrust grows between the man in the street and his neighbour. It can be a question of status, or simple paranoia, but more often it is justified by reality: the injustices of justice to which so many ordinary people are subjected. Over all, the suspicion that one nation has for another hatches the nuclear virus — the cause of our being here today in our various capacities.

Not so long ago a ray of light leapt at me out of the prevailing gloom. A series of historic incidents showed me that suspicion of one nation for another can be allayed if the head of one is brave or idealistic enough to take the lead and give his opponent reason for trust. I'm referring to an article by the American Professor Abraham Keller, President of Educators for Social Responsibilities, in which he points out that:

> On June 10, 1963, in a commencement address at the American University in Washington, President John Kennedy announced that the United States would no longer conduct bomb tests in the atmosphere. His speech was not only printed immediately in *Pravda,* but the *Voice of America* program which broadcast it was not jammed as its programs regularly were and the President's words quickly reached the Russian people. On June 11, in the United Nations, the Soviet Union withdrew its objections to a Western proposal to send observers to war-torn Yemen, a proposal it had been opposing as a capitalist plot. Three days later, on June 14, again at the United Nations, the United States withdrew its objection to the seating of the Hungarian delegation which it had been calling a puppet of the Soviet. The next day, June 15, President Kruschev took to the air congratulating Kennedy on his speech and announced that the Soviet Union was discontinuing the production of strategic bombers. In July the Soviet Union stopped its bomb tests in the atmosphere and on August 5 representatives of the two nations made the test ban final by signing the Treaty of Moscow, which was ratified by the United States Senate in September. On October 9 Kennedy lifted the grain embargo and allowed the shipment of 250-million dollars worth of wheat to the Soviet Union. Also in October the two nations signed a pact agreeing not to orbit nuclear weapons in space. Where years of negotiation for a test ban had failed, a single step by one of the partners brought brilliant success, and, though briefly, established a momentum of goodwill, which went beyond the bomb tests themselves. Then, in November 1963, John Kennedy was assassinated, and hopes for a continuation of the process which he had set in motion, and which had been received with exuberance by many, were buried with him.

I don't know whether many of those Australians who write to the papers and see the Soviet as a permament butt for vilification are aware of this exchange between the two major powers, but I would like them to take notice of Professor Keller's article as evidence that the Soviet, in spite of their brutal treatment of dissidents, the labour camps, and atrocities worthy of the Tsars they replaced, can respond to the civilised approach. Behind the Slav visor and the stereotype diplomatic suits there may even lurk a soul. How else could this barbarous nation have produced the poets Pushkin and Pasternak, the playwright Chekov, the novelists Tolstoy and Dostoevsky, and composers such as Moussorgsky, Tchaikovsky, Shostakovich and Stravinsky. (Incidentally, a backward glance at the barbarous and brutal shows that decent Australians contributed a fair measure of brutality in their treatment of convicts and Aborigines.)

To return to those writers of letters to the daily press, I'd like to quote one from a John McCrae of Balmain, Sydney, who seems to speak with authority instead of the hysteria discernible in many other outraged correspondents. 'In Communist countries peace marchers are shot', asserts one; to which McCrae replies, 'This is not so. On August 1 last year I was in Kiev, and on that day I saw a massive peace and disarmament rally through the city streets. On huge canvas placards suspended above and across the streets, were printed slogans in many foreign languages including English, advocating peace, and denouncing all forms of warfare. At the base of a nearby World War II memorial column, flowers surrounded it knee-deep, a tribute to the fallen.' The soil in which the seeds of truth and trust can be sown is there. It is to be found in the Communist state of East Germany, where a large demonstration was organized in Dresden to commemorate the 37th anniversary of the city's destruction by British and American bombers causing the deaths of 35 000 people — in East Germany where Bishop Hempel has proposed that the state should renounce the stationing of Soviet-built nuclear missiles in its territory, and has called for compulsory 'peace education' in East German schools.

I pray that the words spoken at this conference may carry beyond walls and reach thousands of ears hitherto deaf to warnings of the final catastrophe. I pray that we may convey to them the darkness of night which will fall upon the earth, the death of life in the oceans, the death of crops, trees, and herds, and the immediate or painful lingering death which will come to most of us. I pray that those who hear and see will join those other people of good will who are already working to avert disaster and that they will take heart from the positive results achieved by John Kennedy before his assassination.

Perhaps I can draw your attention to some other examples of non-violent revolution. No need by now to mention the achievements of Gandhi in South Africa and India. But many will be unaware that the Finns between 1901 and 1905 conducted a campaign of non-violent resistance to Russian oppression; this was completely successful and in 1905 the law imposing conscription on the Finns was repealed. Again, the long campaign of non-violent resistance conducted by the Hungarians under Deák was crowned with complete success in 1867. Deák refused political power and personal distinction, was unshakably a pacifist, and without shedding blood compelled the Austrian government to restore the Hungarian constitution. Deák succeeded where the ambitious, power-loving militarist Kóssuth had failed in 1848. In Germany, two campaigns of non-violent resistance were successfully carried out against Bismarck — the Kulturkampf by the Catholics, and the working-class campaign, after 1871, for recognition of the Social Democratic Party. More recently, non-violent resistance and non-co-operation were successfully used in modern Egypt against British domination. A striking example of the way in which even a threat of non-violent non-co-operation can avert war was provided by the British Labour Movement in 1920. The Council of Action formed that year warned the government that if it persisted in its scheme of sending British troops to Poland for an attack on the Russians, a general strike would be called, Labour would refuse to transport munitions or men, and a complete boycott of the war would be declared. Faced by this ultimatum, Lloyd George abandoned his plans for war.

History shows us repeatedly that non-violence can achieve positive results. But — again to quote Aldous Huxley:

> People prepare for war among other reasons because war is in the great tradition, because war is exciting and gives them certain personal and vicarious satisfaction, because their education has left them militaristically minded, because they live in a society where success, however achieved, is worshipped, and where competition seems more natural than co-operation. Hence the general reluctance to embark on constructive policies directed towards the removal at least of the economic causes of war. Hence, too, the extraordinary energy rulers and even the

> ruled put into such war-provoking policies as rearmament, the centralisation of executive power, and the regimentation of the masses.

In Huxley's day, such policies were pursued by the great dictators Mussolini and Hitler. Very disturbing to many of us today is the way Thatcher fanned the emotions of the democratic British to fever pitch and led them into that lamentable swashbuckling expedition to the Falklands, to distract their attention from the state of affairs at home.

In the words of Mussolini, 'Fascism believes that war alone brings up to its highest tension all human energy and puts the stamp of nobility upon the people who have the courage to meet it.' Surely the same sentiments were belted out loud and clear in Thatcher's tirades at the time of the Falklands campaign.

Of course torrents of water have flooded under the bridge since the illuminating months which preceded John Kennedy's assassination, waters which haven't carried us forward, but sucked us back to where we were. The aged cowboy filmstar Ronald Reagan has called upon the scientific community who gave us nuclear weapons to turn their great talents to the cause of mankind and world peace — to give us the means of rendering these nuclear weapons impotent and obsolete. The chief source of stimulus to the fantasies in President Reagan's Reader's-Digest mind is the diabolical Edward Teller, so-called 'Father of the H-Bomb' who is still around at the age of 75. I shall leave Teller's latest nuclear tricks to the experts, confining myself to mention of his obsession that the Russians are totally evil, totally cunning. Nobody is that. I can recognise a certain amount of evil in myself, for instance, but would lay claim to a little good. So with all of us: there is material to work on.

However, I cannot prevent myself suspecting that those who devised the comparatively primitive bombs which were dropped on Hiroshima and Nagasaki, and who have since gone on to develop more sophisticated nuclear weapons, are totally evil. I don't think I am ghoulish in saying that I would like them, and every morally responsible citizen of the world, particularly my fellow Australians of the World War II period, to refresh their memories by referring regularly to the photographic record of the Hiroshima-Nagasaki happening — the rags of human flesh, the suppurating sores, the despair of families blown apart, the disturbed minds, the bleak black gritty plains where the homes of human beings like you and me once stood. Most of all, I would like every Australian *couple* born since Hiroshima and Nagasaki were blasted out of existence to consult these photographic records and for ever after do all in their power to prevent the children they are creating from suffering a fate similar to that thrust upon the children of those two Japanese cities. Let us rouse ourselves and realise this is what we shall have to face. Australians are not prepared for anguish. I don't mean only in the sense of personal bereavement, but in the true spiritual sense, when we feel that God may have forsaken the world, the God many of us probably won't have given a thought to, until the crunch comes in a cosmic flash. Either we are exterminated completely, or worse, we linger on — the rags of flesh, the sores, the disturbed mind. If we are to bear this at all, it will be through God's grace, by cultivating human dignity, and by our ability to dispense with the superfluous details of life as we have known it.

I look to the women of this world, who are in many cases more perceptive than the men and possessed of a determined physical and moral strength — witness the women of Britain and their opposition to cruise missiles.

The other day I came across an aphorism from Rudyard Kipling's *Plain Tales from the Hills.* Not my favourite writer — too much the imperialist bully of his period — but in this little aphorism he reveals a curiously perceptive, *feminine* sensitivity. Listen to it: *A woman's guess is much more accurate than a man's certainty.* From my childhood onward I have felt this particularly in Australia, and contemporary Australian women could play a leading part in preparing us to face and avert nuclear warfare, where all is uncertain, and where the masculine mind may be too orthodox in its approach. I hope I am not castrating anybody by making this remark. I am speaking of women of the stuff from which the early feminists were recruited, before they were persuaded to see themselves as female eunuchs, and surrender much of their strength. As for the male animal, I see him as strengthened by recognising the feminine element in his psyche. The Australian male has shown himself unquestionably courageous in facing up to dangers and death in a series of conventional wars. But the dangers and death we shall have to face in a nuclear war are of a somewhat different order. The fact that the death of a planet may occur raises the issue from a humanist to an eschatological level — spiritual as well as material death, involving judgment, heaven, and hell;

though personally I can't go along with the theological concept of hell. For me, hell is here on earth, living in the shadow of the giant mushroom, with maniacs like Reagan and Teller calling the tune.

That our fate is not entirely in their hands is due to the fact that the people of the world are stirring, finding a voice — enlightened scientists aware of the folly of a war neither side can win — medical men who will bear the brunt of a nuclear disaster without the means for coping with it — high-rank Army officers who have been through other wars and seen the light — economists and sociologists appalled by the billions poured into the manufacture of arms instead of into the bellies of the hungry — churchmen at last recognising their former hierarchic pride and capitalist ambitions. Most encouraging is the stand made by the Roman Catholic bishops of the United States — The Archbishop of Canterbury in Thatcher's dehumanized Britain — and the British Council of Churches, an interdenominational Protestant organisation which, in a recent debate on the nuclear arms race, passed a resolution: 'There are circumstances in which Christian obedience demands civil *dis*-obedience.'

A long course of evil-doing can result in all concerned becoming so sick of destruction and degeneration that they decide to change their ways, thus transforming evil into good. I like to think that the anti-nuclear movement throughout the world is proof that this has begun to happen.

Leaders of the world community have set us a heartening example. How then, can we, the ordinary folk who have no specific role to play, contribute? I include myself in this category because I have no position or skill which might prevent a showdown or alleviate the suffering caused by what we may fail to avert. I am one of you millions of *beings.* Being in itself can be a contribution if it is a concerned being, if we are prepared to offer our *selves* as a sacrifice — that murmurous cluster of human bees I mentioned earlier — a mass sacrifice in the cause of non-violence and the continuance of life on earth.

Curiously, when the fortress of misguided values is occupied by the likes of Reagan and Teller (even members of the British Royal Family have joined the enemy by publicly advocating nuclear deterrents) I have derived immense comfort, hope, faith, inspiration from a great American, the Cistercian monk-teacher-activist Thomas Merton. Initially a contemplative religious,

Merton's spiritual drive was aimed at halting the dehumanization of man in contemporary society, a sickness he saw as leading to mass vilence and ultimately nuclear war. War of any kind is abhorrent. Remember that since the end of World War II, over 40 million people have been killed by conventional weapons. So, if we should succeed in averting nuclear war, we must not let ourselves be sold the alternative of conventional weapons for killing our fellow men. We must cure ourselves of the habit of war. Or is this too fanciful an aim? However the sceptics may shrug, I shall continue to preach non-violence to all those who face the contingency of nuclear or any kind of war, and hope that my fellow Australians, from reading and hearing about murder, rape, arson, petty theft and condoned embezzlement in their everyday life, in this so-called 'pure' country, will not have become so callous that they ignore the greatest opportunity for unity which history has offered the nations of the world. This I see as the positive side of the nuclear threat. The spirit may triumph where politics (the League and the United Nations), socio-political faiths such as Marxism, Italian Fascism and German National-Socialism — all have failed. I see our only hope in faith, charity, and in humbling ourselves before man and God.

In the 14th Century an anonymous English mystic wrote a book called *The Cloud of Unknowing,* the main theme of which is that God cannot be apprehended by man's intellect and that only love can pierce the 'cloud of unknowing' which lies between Him and us. I feel that in my own life anything I have done of possible worth has happened in spite of my gross, worldly self. I have been no more than the vessel used to convey ideas above my intellectual capacities. When people praise passages I have written, more often than not I can genuinely say, 'Did I write that?' I don't think this is due to my having a bad memory, because I have almost total recall of trivialities. I see it as evidence of the part the supernatural plays in lives which would otherwise remain earthbound.

It occured to me in a recent re-reading of *The Cloud of Unknowing* and through my discovery of Tom Merton's work that there may be a connection between the cloud in which God's wisdom is hidden from the human intellect and that other cloud which has never dispersed from above Hiroshima and Nagasaki. It could be that this diabolical mushroom, preserved by photographic plates and human memory, is given us as an icon, or reminder that we contain the seeds of evil and destruction as well as the seeds of divine regeneration. Time is running out. In 1983 it is up to us to choose which we are going to cultivate.

INDEX